COMPUTER PROGRAMS FOR COMPUTATIONAL ASSISTANCE IN THE STUDY OF LINEAR CONTROL THEORY

JAMES L. MELSA
STEPHEN K. JONES

SECOND EDITION

COMPUTER PROGRAMS FOR COMPUTATIONAL ASSISTANCE IN THE STUDY OF LINEAR CONTROL THEORY

SECOND EDITION

JAMES L. MELSA
Professor
Information & Control Sciences Center
Southern Methodist University

STEPHEN K. JONES
Visiting Assistant Professor
Information & Control Sciences Center
Southern Methodist University

196 28

McGRAW-HILL BOOK COMPANY
New York   St. Louis   San Francisco   Düsseldorf
Johannesburg   Kuala Lumpur   London   Mexico
Montreal   New Delhi   Panama   Rio de Janeiro
Singapore   Sydney   Toronto

COMPUTER PROGRAMS FOR COMPUTATIONAL ASSISTANCE IN THE STUDY
OF LINEAR CONTROL THEORY

34567890WHWH7987654

Library of Congress Cataloging in Publication Data

Melsa, James L
   Computer programs for computational assistance
in the study of linear control theory.

   1. Control theory — Computer programs.
I. Jones, Stephen K., joint author.   II. Title.
QA402.3.M386   1973   629.8'32'0285425   72-13701
ISBN 0-07-041498-X

PREFACE

This book is intended to provide assistance in solving computational problems associated with the study and application of linear control theory. A series of thirteen digital computer programs is presented which permit one to carry out the analysis, design and simulation of a broad class of linear control problems. Although the book was written primarily to serve as a supplement to a basic course in linear control, the practicing engineer will also find it to be of considerable value for realistic design and analysis problems.

Extensive use has been made of subprograms in the development of the computer codes so that the material could be easily adapted for use on problems not treated here. All of the subprograms are described in some detail in two appendices in order to further facilitate flexibility. It is hoped that this book and its associated computer codes will not be considered as the final answer to computational problems of linear control but rather the basic essentials from which further developments may be extracted. In this regard, the authors would appreciate receiving any reports on uses or modification of the material contained in this book.

In order to use the codes described in this book, it is necessary to possess only a rudimentary knowledge of FORTRAN coding and an ability to use a card punch and to submit a FORTRAN program. For those readers who are more familiar with FORTRAN, listings of the programs as well as other information have been provided to assist in the understanding and possible modification of the codes. The reader is also assumed to be familiar with the basic concepts of linear control theory as might be obtained from any one of a number of available basic control texts.

The computer codes described in this book have been developed over a number of years and have been used by the authors and others to solve a wide variety of both academic and practical problems. Some typical applications are discussed in Chapter 4. Nevertheless, one can never completely test any program especially on all available FORTRAN compilers. The authors apologize for any problems which may arise and would appreciate information on any special difficulties encountered.

In order to facilitate the use of the computer codes described in this book, a prepunched deck of cards, which includes all programs, subprograms and example input, may be obtained by writing to James L. Melsa, Information and Control Sciences Center, Southern Methodist University, Dallas, Texas 75222. The programs are all written in basic FORTRAN IV language and use the IBM029 punch format.

The authors express their appreciation to many students and colleagues who assisted in this project. The inspiring leadership of Dr. Andrew P. Sage, Director of the Information and Control Sciences Center, SMU, and Dr. Thomas L. Martin, Dean of the Institute of Technology, SMU, has been a source of constant encouragement.

CONTENTS

1 INTRODUCTION

1.1 PURPOSE AND OUTLINE

In the study of linear control theory, one often encounters computations which, although theoretically simple, can be an impediment to the learning of the basic concepts. For example after completing a control system design, it may be very helpful to the educational process to evaluate this design. The labor involved in the determination of accurate frequency response or root locus plots or determining the step response of the system may be prohibitive. In addition, it is often desirable to study high-order realistic design problems in order to emphasize the practical usefulness of the theory. Without a method of computational assistance, such problems are usually avoided because the computational effort required far exceeds the benefits obtained.

The goal of this book is to provide a set of computer codes which solve almost all of the computational problems of linear control theory. These codes allow one to study realistic problems in detail without extensive computation effort. In addition, the use of these codes give experience with the concepts of computer aided design and analysis which are so popular in current engineering practice.

It is assumed that the reader is familiar with the basic concepts of linear control theory as may be obtained from any one of a number of available texts. Extensive attention is given to the state variable description of linear control systems and the use of linear state variable feedback. The interested reader will find a detailed treatment of this material in the book: Linear Control Systems.[1] The notation and system formulation used in this book follows closely that used in Linear Control Systems.

In order to use the material in this book, it is not required that the reader possess any knowledge of numerical methods. Only the very basic essentials of the FORTRAN language and some simple knowledge of computer card deck preparation are needed. For the reader who is familiar with FORTRAN, listings of the programs and subprograms as well as notes on coding are presented for reference.

The programs are divided into two broad classes: (1) State variable programs and (2) Transfer function programs. The five state variable programs described in Chapter 2 allow one to obtain certain matrix functions such as the inverse, determinant, resolvent matrix and state transition matrix as well as to analyze and design linear state variable feedback systems. Two programs are presented which compute the time response of linear state variable feedback systems. There is a program for studying the sensitivity of closed-loop systems to parameter variations and six programs for the design of linear systems.

---

[1]J. L. Melsa and D. G. Schultz: Linear Control Systems, McGraw-Hill Book Co., Inc., 1969.

Three transfer function programs are discussed in Chapter 3. These programs may be used to determine the frequency response and root locus of a system. The third transfer function program is used to obtain the partial fraction expansion of a rational function. Some typical applications of the computer programs are presented in Chapter 4.

All of the subprograms used are described in Appendix A. These subprogram descriptions are included to assist the interested reader in understanding the operation of the programs and also to permit the reader to generate new programs from these subprograms. Extensive use was made of subprograms in the development of these computer codes in order to facilitate such flexibility. For example one might wish to combine the sensitivity analysis and time response programs or the state variable feedback and root locus programs in order to carry out some particular study.

Appendix B presents a set of three graphical display subprograms which are used in the two preceding chapters. These display programs are quite general and allow one to obtain a quick graphical representation of the results of the programs on a line printer. The subprograms discussed are logarithmic plots, X - Y plot and a X - versus time plot.

Each of the thirteen program write-ups presented in Chapters 2 and 3 follow the same format. After a brief statement of the purpose of the program, the theoretical concepts involved in the development of the computer codes are discussed. The input and output format of the program is explained and then illustrated by means of an example. The input data deck is shown and the resulting program output is presented and discussed. In some cases it has been necessary to slightly modify or condense the computer output in order to fit it within page-size limitations. The output should however remain indicative of the type of output which will be obtained. Finally, the listing of each program is presented for reference.

In general, the programs have been limited to tenth-order problems. Although there is no basic limitation in the algorithms which make this constraint necessary, it was felt that almost all academic and most practical problems satisfy this limitation. If necessary, this limitation may be removed by extending the appropriate dimension statements. However it is not recommended that problems of excessively high order be treated since no major effort has been directed to a study of the efficiency of the numerical methods used. For an excellent discussion of numerical methods, the interested reader is directed to the book: Numerical Methods for Scientists and Engineers.[1]

1.2 INPUT FORMAT

The input format for each program is described in detail in Chapters 2 and 3; the purpose of this section is to comment on certain general features of the input formats which applied to all of the programs. A primary consideration in the selection of the input format was to make the use of the programs as simple as possible while still retaining sufficient flexibility. A second consideration was to make the input formats of the various programs as nearly uniform as possible. Because of these considerations, the input formats selected often contain more cards than necessary to provide the program with input data. On the other hand the input formats are easy to remember since a large majority of the cards have identical forms and the same general format is used in every program. In addition the input data deck is designed so as to be closely related to the basic problem statement so that it is easy to make modifications in the problem if desired.

---

[1] R. W. Hamming: Numerical Methods for Scientists and Engineers, McGraw-Hill Book Co., Inc., 1962.

All of the programs can be used to solve more than one problem in a single run by simply placing the various input decks one after another. In other words, an input deck is prepared for each problem and these decks are then added together to form a composite input deck for the program.

In the description of the input data format, the FORTRAN format type information is included for the reader who is familiar with the FORTRAN language. The first card of every input deck contains problem identification information which may be used for later reference. Any desired alphanumeric characters may be placed in the first twenty columns of this first card. If appropriate, the next two columns (21-22) of the first data card will contain information concerning the problem order in fixed-point format. The reader is reminded that fixed-point numbers are always right justified in their fields.

The remaining input cards will vary depending on the particular program under consideration. However there are some general format rules which are followed for entering information in the form of matrices and in terms of polynomials. Let us consider each of these types of input separately.

1.2-1 *MATRIX DATA*   Only two types of matrices are considered in the programs described in this book: square matrices and column vectors. The elements of a square matrix are always read one row at a time much as they appear in the original array. The numbers are placed in ten-column fields in floating-point format. If the dimension N of the matrix is greater than eight, the elements continue on the next card. The elements of the next row, however, start on a new card. Therefore if $N \le 8$, N cards are used to represent the matrix while if $8 < N \le 10$, 2N cards are necessary. Suppose for example that we wish to enter the 3 by 3 matrix

$$\underset{\sim}{A} = \begin{bmatrix} 3 & 0 & 1 \\ -1 & 2 & 0 \\ 4 & 1.2 & 3 \end{bmatrix}$$

then the input cards would take the form shown in Table 1.2-1. Note that the decimal point should be punched since a floating point format is being used.

Table 1.2-1

Input Card for a 3 by 3 matrix

Card No.	Column No.				
1	3.0	0.0	1.0		
2	-1.0	2.0	0.0		
3	4.0	1.2	3.0		

In the case of column vectors, the elements are read in <u>column</u> form in order to minimize the number of input cards. In other words the elements are read as if the vector were a row vector. If the dimension N of the vector is less than or equal to eight, only one card will be used to input the vector; if $8 < N \le 10$, then two cards will be needed. In the input format descriptions in Chapters 2 and 3, it has been assumed that $N \le 8$ in order to simplify the discussion. Whenever $N > 8$, it is simply necessary to double the number of cards used to describe any vector or square matrix.

1.2-2 *POLYNOMIAL DATA* Polynomials may be entered in two different formats. All programs which accept polynomial inputs allow either format to be used. The polynomial may be entered as polynomial coefficients or as polynomial factors; these two formats will be referred to as P mode or F mode respectively.

If the P mode is selected the coefficients of the polynomials are entered in ten-column, floating-point fields just as if they were elements of a matrix. The constant term is given first and the coefficient of the highest term is assumed to be unity. This coefficient is read in although it is set to unity and not used. Note that if the degree of the polynomial is equal to eight it is necessary to add a second card which may be blank if desired. In other words, if an eighth order polynomial is to be read, the first eight coefficients are entered in the eight ten-column fields on one card, the coefficient of s^8 must then be placed on the next card even though it must be unity.

If it is desired to enter the polynomial in F mode as polynomial factors, then one factor is placed on each card. The real part of the factor is entered in the first ten columns and the imaginary part is in the next ten columns. The real part is <u>positive</u> is the factor is in the <u>left</u> <u>half</u> <u>plane</u>. For complex conjugate roots only one card is used; the program will supply the complex conjugate. The imaginary part will <u>always</u> be positive.

In order to indicate which of the two polynomial modes is to be used, a control card is placed in front of each polynomial input. A P or F is entered in the first column of this card to indicate which mode is being used and the degree of the polynomial is entered in fixed-point form in the next two columns. Let us consider the third-order polynomial P(s) given by

$$P(s) = 5 + 9s + 5s^2 + s^3 = (s + 1)[(s + 2)^2 + 1]$$

The input deck for this polynomial in P mode is shown in Table 1.2-2a and in F mode in Table 1.2-2b. Note that the coefficient of s^3 is assumed to be unity and need not be entered.

Table 1.2-2

Polynomial Input Deck: (a) P mode, (b) F mode.

(a)

(b)

1.3 OUTPUT FORMAT

The output of all of the programs is almost self explanatory. Only a few comments are indicated here to clarify any particular factors which might cause confusion. At the beginning of every program output, the problem identification supplied by the user and the name of the program is printed for reference. The input data is also listed for reference and finally the actual output of the program is presented. If the output has several parts, the problem identification information is repeated for each part. In the presentation of the actual program output for the example problems discussed in Chapters 2 and 3, certain modifications and condensations have been necessitated by the page size limitation. However the basic content and nature of the output should be unchanged. It is suggested that the reader prepare the input for and run the example problems for himself.

In the program output, square matrices are always written as they would normally appear with rows horizontally and columns vertically. Column vectors are printed in transposed form across the page just as they are read in.

Polynomials are given in both factored and unfactored form independent of which input form was used. The coefficients are given in ascending order with the constant term first. The factors are listed in the original form with negative real part indicating that the root is in the left half plane. The reader should note that there is a sign change between the input format for a factor and the way that it is listed.

2 STATE VARIABLE PROGRAMS

2.1 INTRODUCTION

In this chapter we discuss a group of ten programs which may be used for
the analysis and design of linear control systems represented in state variable
forms as

$$\dot{\underset{\sim}{x}}(t) = \underset{\sim}{A}x(t) + \underset{\sim}{b}u(t)$$

$$u(t) = K[r(t) - \underset{\sim}{k}^T \underset{\sim}{x}(t)]$$

$$y(t) = \underset{\sim}{c}^T \underset{\sim}{x}(t)$$

The BASIC MATRIX (BASMAT) program, discussed in Sec. 2.2, allows one to
compute the determinant, inverse, characteristic polynomial and eigenvalues of
a square matrix $\underset{\sim}{A}$. In addition BASMAT may be used to compute the resolvent
matrix $(s\underset{\sim}{I} - \underset{\sim}{A})^{-1}$ and state transition matrix $\exp(\underset{\sim}{A}t)$.

Two programs are provided for determining the time response of the linear
feedback control system above. The RATIONAL TIME RESPONSE (RTRESP) program,
Sec. 2.3, requires that the input function r(t) have a rational time response
and that there be no repeated eigenvalues in the combination of the system and
input. The RTRESP program gives a closed-form expression for the time response.
The GRAPHICAL TIME RESPONSE (GTRESP) program of Sec. 2.4, is used to determine
and graphically display the time response for an arbitrary input. Note that
both of these programs may be used to study open-loop systems by letting K = 0
and unforced systems by setting r(t) = 0.

One of the basic problems of linear system analysis and design is the
question of sensitivity to parameter variations. The SENSITIVITY ANALYSIS
(SENSIT) program may be used to provide some information on sensitivity. The
SENSIT program, which is discussed in Sec. 2.5, permits one to study the effect
on the closed-loop poles of changes in elements of $\underset{\sim}{A}$, $\underset{\sim}{b}$, $\underset{\sim}{k}$ or K. A graphical
output is provided to assist in the evaluation.

The next section of this chapter discusses the STATE VARIABLE FEEDBACK
(STVARFDBK) program. This program has been widely used in both academic and
industrial environments for the analysis and particularly the design of linear
state variable feedback control systems. The STVARFDBK program may be used to
find both the open-loop plant and closed-loop system transfer functions and can
in addition design closed-loop system from desired closed-loop transfer function
specifications.

Sections 2.7 to 2.9 present three programs which may be used to design
state variable feedback systems with inaccessible state variables. The program
OBSERV is used to find the observability index of a system with determines
the order of the compensator needed for either the Luenberger observer program
(LUEN) or the series compensation program (SERCOM). The last two sections
present two programs which may be used for designing state variable feedback
structures for multiple-input, multi-output systems.

6

2.2 BASIC MATRIX (BASMAT)

Given a matrix A, the BASIC MATRIX program can compute the determinant of A, det A, the inverse of A, A^{-1}, the characteristic polynomial det$(sI - A)$ and eigenvalues of A, λ_i as well as the state transition matrix $\Phi(t) = \exp(At)$ and the resolvent matrix $\Phi(s) = (sI - A)^{-1}$.

2.2-1 *THEORY* The BASMAT program reads the elements of the matrix A row by row and an option card which indicates what functions of A are desired. The main program of BASMAT then calls subprograms to perform the appropriate calculations, these subprograms are discussed in Appendix A.

2.2-2 *INPUT FORMAT* The input data for the BASIC MATRIX programs consists of the usual identification card which also contains the dimension N of the matrix A in columns 21-22. The elements of A are then read row by row from the next N cards[1]. The last data card determines which computations are to be performed by BASMAT. If a column is zero or blank, the related operation is performed; if the column contains any non-zero integer (1 to 9), then the related operation is suppressed. If the last card is completely blank, all operations are performed. The input format is summarized in Table 2.2-1 for easy reference.

Table 2.2-1

Input Format for BASIC MATRIX

Card Number	Column Number	Description	Format
1	1-20 21-22	Problem identification N = dimension of $A \leq 10$	5A4, I2
2	1-10 11-20 etc.	a_{11} a_{12} ...	8E10.5
3	1-10 11-20 etc.	a_{21} a_{22} ...	8E10.5

N + 2	1
	2
	3
	4
	
	5
	6
		IDET $\neq 0$ suppress determinant
		INV $\neq 0$ suppress inverse
		NRM $\neq 0$ suppress resolvent
		ICP $\neq 0$ suppress characteristic
		   polynomial
		IEIG $\neq 0$ suppress eigenvalues
		ISTM $\neq 0$ suppress state transition
		   matrix
			6I1

---

[1] N is assumed to be less than or equal to 8 for simplicity. See Sec. 1.2.

2.2-3 *OUTPUT FORMAT* After the problem identification, the matrix $\underline{A}$ is printed for reference. Then depending on the options selected, $\det(\underline{A})$, $\underline{A}^{-1}$, $\underline{\Phi}(s)$, the characteristic polynomial, eigenvalues, and $\underline{\Phi}(t)$ are printed. The resolvent matrix is written as $\underline{\Phi}(s) = adj(s\underline{I}-\underline{A})/\det(s\underline{I}-\underline{A})$ and the matrix numerator polynomial $adj(s\underline{I}-\underline{A})$ is expressed as matrix coefficients of powers of s so that it takes the form

$$adj(s\underline{I}-\underline{A}) = \underline{F}_1 + \underline{F}_2 s + \ldots + \underline{F}_N s^{N-1}$$

The state transition matrix is expressed in a form similar to the $adj(s\underline{I}-\underline{A})$ as matrix coefficients times the natural modes $\exp(\lambda_i t)$; complex eigenvalues are written as damped sine and cosine terms.

2.2-4 *EXAMPLE* In order to illustrate the use of the BASMAT program, let us consider the following matrix

$$\underline{A} = \begin{bmatrix} 0 & 1 & 0 \\ 0 & 0 & 1 \\ -2 & -3 & -3 \end{bmatrix}$$

We will use all of the options available for this example so the last card is left blank. The form of the input data cards for this example is shown in Table 2.2-2.

Table 2.2-2
Input Data Deck for BASMAT Example

Card No.	Column No. 5	10	15	20	25	30
1	BASMAT EXAMPLE	3		3		
2	0.0	1.0		0.0		
3	0.0	0.0		1.0		
4	-2.0	-3.0		-3.0		
5	(BLANK CARD)					

The program output for this example problem is shown in Table 2.2-3. From this output, we see that $\phi_{11}(s)$ is

$$\phi_{11}(s) = \frac{3 + 3s + s^2}{2 + 3s + 3s^2 + s^3}$$

and the corresponding element of the state transition matrix is

$$\phi_{11}(t) = 0.667e^{-0.5t} \cos 0.866t + 1.15e^{-0.5t} \sin 0.866t + 0.333e^{-2t}$$

8

Table 2.2-3

Program Output for BASMAT Example

```
BASIC MATRIX PROGRAM
PROBLEM IDENTIFICATION:      BASMAT EXAMPLE 3
```

****

THE A MATRIX

```
 0.0                  1.0000000E 00        0.0
 0.0                  0.0                  1.0000000E 00
-2.0000000E 00       -3.0000000E 00       -3.0000000E 00
```

THE DETERMINANT OF THE MATRIX

-2.0000000E 00

THE INVERSE OF THE MATRIX

```
-1.5000000E 00       -1.5000000E 00       -5.00C0000E-01
 1.0000000E 00        0.0                  0.0
 0.0                  1.0000000E 00        0.0
```

****

THE MATRIX COEFFICIENTS OF THE NUMERATOR OF THE RESOLVENT MATRIX

THE MATRIX COEFFICIENT OF S**2

```
1.0000000E 00         0.0                  0.0
0.0                   1.0000000E 00        0.0
0.0                   0.0                  1.0000000E 00
```

THE MATRIX COEFFICIENT OF S**1

```
 3.0000000E 00        1.0000000E 00        0.0
 0.0                  3.0000000E 00        1.0000000E 00
-2.0000000E 00       -3.0000000E 00        0.0
```

THE MATRIX COEFFICIENT OF S**0

```
 3.0000000E 00        3.0000000E 00        1.0000000E 00
-2.0000000E 00        0.0                  0.0
 0.0                 -2.0000000CE 00       0.0
```

****

THE CHARACTERISTIC POLYNOMIAL — IN ASCENDING POWERS OF S

```
1.9999990E 00         3.C000000E 00        3.0000000E 00           1.0
```

****

```
THE EIGENVALUES OF THE A MATRIX
    REAL PART          IMAGINARY PART

-5.0000048E-01       -8.6602509E-01
-5.0000048E-01        8.6602509E-01
-1.9999990E 00        0.0
```

9

Table 2.2-3 (cont.)

****

THE ELEMENTS OF THE STATE TRANSITION MATRIX

THE MATRIX COEFFICIENT OF EXP(-5.000005E-01)T*COS(8.660251E-01)

```
  6.6666591E-01        -3.3333445E-01        -3.3333409E-01
  6.6666842E-01         1.6666679E 00         6.6666794E-01
 -1.3333349E 00        -1.3333359E 00        -3.3333492E-01
```

THE MATRIX COEFFICIENT OF EXP(-5.000005E-01)T*SIN(8.660251E-01)

```
  1.1547022E 00         1.7320518E 00         5.7735133E-01
 -1.1547022E 00        -5.7735050E-01        -3.5762787E-07
  0.0                  -1.1547031E 00        -5.7735038E-01
```

THE MATRIX COEFFICIENT OF EXP(-1.999999E 00)T

```
  3.3333433E-01         3.3333457E-01         3.3333439E-01
 -6.6666853E-01        -6.6666758E-01        -6.6666794E-01
  1.3333340E 00         1.3333340E 00         1.3333340E 00
```

2.2-5   *SUBPROGRAMS USED*   The following subprograms are used by this program:

(1)   DET
(2)   CHREQ
(3)   PROOT
(4)   SIMEQ
(5)   STMST

Please see Appendix A for a description of each of these subprograms.

2.2-6   *PROGRAM LISTING*

```
C   BASIC MATRIX PROGRAM
C   SUBPROGRAMS USED:   CHREQ, SIMEQ, STMST, PROOT, DET.
        DIMENSION A(10,10),EIGR(10),EIGI(10),C(11),AINV(10,10),
      *     NAME(5)
 2001 FORMAT (5A4,I2)
 2002 FORMAT (8E10.5)
 2003 FORMAT (1P6E20.7)
 2004 FORMAT (1H0,5X,16HTHE A MATRIX      ,/)
 2005 FORMAT (1H0,5X,32HTHE CHARACTERISTIC POLYNOMIAL -
      *     24HIN ASCENDING POWERS OF S   /)
 2006 FORMAT (1H0  ,5X,31HTHE EIGENVALUES OF THE A MATRIX)
 2007 FORMAT (9X,9HREAL PART,8X,14HIMAGINARY PART,/)
 2008 FORMAT (1H1,5X,20HBASIC MATRIX PROGRAM)
 2009 FORMAT (6X,23HPROBLEM IDENTIFICATION:,5X,5A4)
 2010 FORMAT (1H0,5X,29HTHE DETERMINANT OF THE MATRIX/)
 2011 FORMAT (1H0,5X,25HTHE INVERSE OF THE MATRIX/)
 2012 FORMAT (1H0,45(1H*))
 2013 FORMAT (6I1)
    4 READ (5,2001,END=10) (NAME(I),I=1,5),N
      DO 1 I=1,N
    1 READ 2002, (A(I,K),K=1,N)
      READ 2013, IDET,INV,NRM,ICP,IEIG,ISTM
      PRINT 2008
```

```
      PRINT 2009, (NAME(I),I=1,5)
      PRINT 2012
      PRINT 2004
      DO 2 I=1,N
   2  PRINT 2003, (A(I,K),K=1,N)
      IF (IDET.NE.0) GO TO 5
      D=DET(A,N)
      PRINT 2010
      PRINT 2003, D
   5  IF (INV.NE.0) GO TO 15
      PRINT 2011
      CALL SIMEQ(A,C,N,AINV,C,IERR)
      IF (IERR.EQ.0) GO TO 15
      DO 20 I=1,N
  20  PRINT 2003, (AINV(I,J),J=1,N)
  15  CALL CHREQ(A,N,C,NRM)
      CALL PROOT(N,C,EIGR,EIGI,+1)
      IF (ICP.NE.0) GO TO 30
      PRINT 2012
      PRINT 2005
      NN=N+1
      PRINT 2003, (C(I),I=1,NN)
  30  IF (IEIG.NE.0) GO TO 35
      PRINT 2012
      PRINT 2006
      PRINT 2007
      DO 3 I=1,N
   3  PRINT 2003, EIGR(I),EIGI(I)
  35  IF (ISTM.NE.0) GO TO 25
      CALL STMST(N,A,EIGR,EIGI,ISTM)
  25  GO TO 4
  10  STOP
      END
```

2.3 RATIONAL TIME RESPONSE (RTRESP)

The RTRESP program determines the time response in closed form of the closed-loop system

$$\dot{\underline{x}}(t) = \underline{A}\underline{x}(t) + \underline{b}u(t)$$

$$u(t) = K[r(t) - \underline{k}^T\underline{x}(t)]$$

$$y(t) = \underline{c}^T\underline{x}(t)$$

due to any initial conditions $\underline{x}(0)$ and input $r(t)$, $t \geq 0$ which has a rational Laplace transform $R(s)$ with a pole-zero excess of at least one.

2.3-1 *THEORY*   Let the Laplace transform of $r(t)$ be given by

$$\mathcal{L}\{r(t)\} = R(s) = K_r \frac{N(s)}{D(s)}$$

where

$$N(s) = n_1 + n_2 s + \ldots + n_k s^{k-1} + s^k$$

$$D(s) = d_1 + d_2 s + \ldots + d_m s^{m-1} + s^m$$

and $m > k \geq 0$. The approach which we shall use will be to form an mth-order dynamic system whose initial condition response (for a specific set of initial conditions) is equal to $r(t)$. Then we may combine this new system with the original system and find the complete response in closed form by the use of STMST (See Appendix A). Let us use phase variables to represent this new system in the form

$$\dot{\underline{x}}_r(t) = \underline{A}_r\underline{x}_r(t)$$

$$y_r(t) = \underline{c}_r^T\underline{x}_r(t)$$

where

$$\underline{A}_r = \begin{bmatrix} 0 & 1 & 0 & \ldots & 0 \\ 0 & 0 & 1 & \ldots & 0 \\ \multicolumn{5}{c}{\cdots\cdots\cdots\cdots\cdots\cdots} \\ 0 & 0 & 0 & \ldots & 1 \\ -d_1 & -d_2 & -d_3 & \ldots & -d_m \end{bmatrix}$$

and

$$\underline{c}_r = \text{col}(K_r n_1, \; K_r n_2, \; \ldots, \; K_r, \; 0, \; \ldots, \; 0)$$

If we let the initial condition of the system be $\underline{x}_r(0) = \text{col}(0, 0, \ldots, 0, 1)$, then the time response of the output $y_r(t)$ will be identical to $r(t)$.

Our original closed-loop system can be represented as

$$\dot{\underline{x}}(t) = \underline{A}_k\underline{x}(t) + K\underline{b}r(t)$$

where $\underline{A}_k = \underline{A} - K\underline{b}\underline{k}^T$. Since $y_r(t)$ is equal to $r(t)$ if the initial condition $\underline{x}_r(0)$ is properly chosen, the closed-loop system can also be written as

$$\dot{\underline{x}}(t) = \underline{A}_k\underline{x}(t) + K\underline{b}\underline{c}_r^T\underline{x}_r(t)$$

Now we form the $n + m^{th}$ order augmented system which is unforced

12

$$\dot{\underline{x}}_a(t) = \underline{A}_a \underline{x}_a(t)$$

$$y(t) = \underline{c}_a^T \underline{x}_a(t)$$

where

$$\underline{x}_a(t) = \begin{bmatrix} \underline{x}(t) \\ \hline \underline{x}_r(t) \end{bmatrix} \qquad\qquad \underline{c}_a = \begin{bmatrix} \underline{c} \\ \hline 0 \end{bmatrix}$$

and

$$\underline{A}_a = \begin{bmatrix} \underline{A}_k & | & K\underline{b}\underline{c}_r^T \\ \hline 0 & | & \underline{A}_r \end{bmatrix}$$

If we let the initial condition of this augmented system be

$$\underline{x}_a^T(0) = [\underline{x}^T(0) \;|\; 0, 0, \ldots, 0, 1]$$

then the time response of $\underline{x}(t)$ and $y(t)$ in the augmented system is identical to the forced response of the original system. The time response of the augmented system is given by

$$\underline{x}_a(t) = \underline{\Phi}_a(t)\underline{x}_a(0)$$

where $\underline{\Phi}_a(t)$ is the state transition matrix $\exp(\underline{A}_a t)$ for the augmented system and may be computed by the use of the STMST subroutine (see Appendix A). Note that only the upper left n by n segment of $\underline{\Phi}_a(t)$ is used since we are interested only in the time response of the original system and not the state $\underline{x}_r(t)$.

2.3-2 *INPUT FORMAT* The input data for this program can be divided into two parts: the description of the basic closed-loop system, i.e. $\underline{A}$, $\underline{b}$, $\underline{c}$, $\underline{k}$, and K and the description of the input function r(t) and the initial condition $\underline{x}(0)$. The input format is summarized in Table 2.3-1 for easy reference.

The description is presented in the usual fashion. The first data card consists of problem identification and the order of the system. The next n cards give the $\underline{A}$ matrix while the following four cards define $\underline{b}$, $\underline{c}$, $\underline{k}$, and K respectively. The initial condition vector $\underline{x}(0)$ is given on the next card in the usual row vector form.

The input function r(t) is described by its rational Laplace transform R(s). The numerator and denominator polynomials N(s) and D(s) may be supplied in the form of either polynomial coefficients or roots. For simplicity, we have shown N(s) supplied as polynomial coefficients and D(s) as polynomial roots in Table 2.3-1. Without loss of generality, we have assumed that the coefficient of s^m in D(s) is unity. Note that the coefficient of the highest order term in both the numerator and denominator polynomials are unity and need not be supplied. As usual roots are read in with positive real part if they lie in the left half plane and only one root of a complex conjugate pair is needed.

In this program it is necessary that the total order of the system plus the input, that is n + m, be less than 10. If only an initial condition response is desired, then K_r may be set to zero and any arbitrary polynomials supplied for N(s) and D(s). Because the STMST subroutine does not handle repeated eigenvalues, it is necessary that the combined system and input have no repeated eigenvalues.

13

Table 2.3-1

Input Format for RTRESP

Card Number	Column Number	Description	Format
1	1-20	Problem identification	5A4,I2
	21-22	n = order of the system	
2	1-10	a_{11} = plant matrix	8F10.2
	11-20	a_{12}	
	etc.	...	
3	1-10	a_{21}	8F10.2
	11-20	a_{22}	
	etc.	...	
n+2	1-10	b_1 = control vector	8F10.2
	11-21	b_2	
	etc.	...	
n+3	1-10	c_1 = output vector	8F10.2
	11-21	c_2	
	etc.	...	
n+4	1-10	k_1 = feedback coefficients	8F10.2
	11-21	k_2	
	etc.	...	
n+5	1-10	K = controller gain	8F10.2
n+6	1-10	$x_1(0)$ = initial condition	8F10.2
	11-20	$x_2(0)$	
	etc.	...	
n+7	1-10	K_r = input gain	8F10.2
n+8	1	KEY = P (coefficients supplied)[1]	A1,I2
	2-3	k = numerator polynomial order, $k < m$	
n+9	1-10	n_1 = numerator coefficients	8F10.2
	11-20	n_2	
	etc.	...	
n+10	1	KEY = F (polynomial factors supplied)[1]	A1,I2
	2-3	m = denominator polynomial order	

[1]Both N(s) and D(s) may be supplied in either polynomial coefficient or root form; see text.

14

Table 2.3-1 (cont.)

Card Number	Column Number	Description	Format
n+11	1-10	real part of the first root of D(s)	8F10.2
	11-20	imaginary part of the first root of D(s)	
n+12	1-10	real part of the second root of D(s)	8F10.2
	11-20	imaginary part of the second root of D(s)	
etc.	...	...	

2.3-3 *OUTPUT FORMAT*   After the problem identification, the $\underset{\sim}{A}$, $\underset{\sim}{b}$, $\underset{\sim}{c}$, $\underset{\sim}{k}$ and K values are printed for reference. The initial condition vector $\underset{\sim}{x}(0)$ and the input function R(s) are then presented. The polynomials N(s) and D(s) are presented in both factored and unfactored form independent of which form was used for input.

The time response of $\underset{\sim}{x}(t)$ is presented in the form of vector coefficients of the various natural modes of the system and the input function as

$$\underset{\sim}{x}(t) = \underset{\sim}{\alpha}_1 e^{\lambda_1 t} + \underset{\sim}{\alpha}_2 e^{\lambda_2 t} + \ldots + \underset{\sim}{\alpha}_{m+n} e^{\lambda_{m+n} t}$$

The system output y(t) is presented in the form of scalar coefficients of the same natural modes.

2.3-4 *EXAMPLE*   We wish to determine the time response of the closed-loop system

$$\dot{\underset{\sim}{x}}(t) = \begin{bmatrix} 0 & 1 \\ -1 & -1 \end{bmatrix} \underset{\sim}{x}(t) + \begin{bmatrix} 0 \\ 1 \end{bmatrix} u(t)$$

$$u(t) = 5\{r(t) - [1.0 \quad 0.5]\underset{\sim}{x}(t)\}$$

$$y(t) = [1 \quad 1]\underset{\sim}{x}(t)$$

if the input function is given by

$$R(s) = 2.0 \frac{1}{s}$$

and the initial conditions are $\underset{\sim}{x}(0) = \underset{\sim}{0}$. The input deck for this problem is given in Table 2.3-2 and the resulting program output is given in Table 2.3-3.

Table 2.3-2

Input Data Deck for RTRESP Example

Card No.	Column No.					
	5	10	15	20	25	30
1	RTRESP EXAMPLE			2		
2	0.0	1.0				
3	-1.0	-1.0				
4	0.0	1.0				
5	1.0	1.0				
6	1.0	0.5				
7	5.0					
8	0.0	0.0				
9	2.0					
10	P 0					
11	1.0					
12	F 1					
13	0.0					

Table 2.3-3

Program Output for RTRESP Example

RATIONAL TIME RESPONSE
PROBLEM IDENTIFICATION — RTRESP EXAMPLE

****

THE A MATRIX

0.0   1.000000E 00

-1.000000E 00   -1.000000E 00

THE B MATRIX

0.0   1.000000E 00

THE C MATRIX

1.000000E 00   1.000000E 00

FEEDBACK COEFF.

1.000000E 00   5.000000E-01

Table 2.3-3 (cont.)

GAIN = 5.0000000E 00

:**

INITIAL CONDITIONS - X(0)

0.0   0.0

RGAIN = 2.0000000E 00

NUMERATOR POLYNOMIAL OF R(S) - ASCENDING POWERS OF S

1.000000E 00

DENOMINATOR POLYNOMIAL OF R(S) - ASCENDING POWERS OF S

0.0   1.000000E 00

DENOMINATOR ROOTS ARE
  REAL PART   IMAG. PART

-0.0   0.0

****

THE TIME RESPONSE OF THE STATE X(T)

THE VECTOR COEFFICIENT OF EXP(-1.750000E 00)T*COS(1.713913E 00)T

-1.666666E 00   0.0

THE VECTOR COEFFICIENT OF EXP(-1.750000E 00)T*SIN(1.713913E 00)T

-1.701758E 00   5.834599E 00

THE VECTOR COEFFICIENT OF EXP(-0.0   )T

1.666666E 00   1.907349E-06

****

THE TIME RESPONSE OF THE OUTPUT Y(T)

THE COEFFICIENT OF EXP(-1.750000E 00)T*COS(1.713913E 00)T

-1.666666E 00

THE COEFFICIENT OF EXP(-1.750000E 00)T*SIN(1.713913E 00)T

4.132840E 00

THE COEFFICIENT OF EXP(-0.0   )T

1.666668E 00

****

From Table 2.3-3, we see that the time response for $x_1(t)$ is

$$x_1(t) = +1.67 - 1.67e^{-1.75t} \cos 1.71t - 1.70e^{-1.75} \sin 1.71t$$

while the output is

$$y(t) = 1.67 - 1.67e^{-1.75t} \cos 1.71t + 4.13e^{-1.75t} \sin 1.71t$$

2.3-5 *SUBPROGRAMS USED*   The following subprograms are used by this program:

      (1)   CHREQA
      (2)   PROOT
      (3)   SEMBL
      (4)   STMST

See Appendix A for a description of each of these subprograms.

2.3-6 *PROGRAM LISTING*

```
C   RATIONAL TIME RESPONSE (RTRESP)
C   SUBPROGRAMS USED: CHREQA,PROOT,SEMBL,STMST.
        COMMON CHI(10,10,10)
        DIMENSION A(10,10),B(10),FDBG(10),RNUM(10),RDEN(10),
      *          PSI(10,10),C(10),COEFF(10),EIGR(10),EIGI(10),
      *          XO(10),XPM(10),D(10),NAME(5),RTR(10),RTI(10),
      *          CS(11)
        DATA IPP/1HP/
 1000 FORMAT (5A4,I2)
 1001 FORMAT (8F10.2)
 1002 FORMAT (1H0,5X,9(1PE13.6,4X))
 1003 FORMAT (1H0,  5X,25HTHE VECTOR COEFFICIENT OF
      *   5H EXP(,1PE13.6,7H)T*COS(,1PE12.6,2H)T)
 1004 FORMAT (1H0,  5X,25HTHE VECTOR COEFFICIENT OF,
      *   5H EXP(,1PE13.6,7H)T*SIN(,1PE12.6,2H)T)
 1005 FORMAT (1H0,5X,25HTHE VECTOR COEFFICIENT OF,
      *   5H EXP(,1PE13.6,2H)T)
 1006 FORMAT(1H1,3X,24H RATIONAL TIME RESPONSE   /
      *   5X,25HPROBLEM IDENTIFICATION -   ,5A4)
 1007 FORMAT (1H0,5X,13HTHE A MATRIX  /)
 1008 FORMAT (1H0,5X,13HTHE B MATRIX  /)
 1009 FORMAT (1H0,5X,13HTHE C MATRIX  /)
 1010 FORMAT (1H0,5X,15HFEEDBACK COEFF.  /)
 1011 FORMAT (1H0,5X,26HINITIAL CONDITIONS - X(0)   /)
 1012 FORMAT (1H0,5X,28HNUMERATOR POLYNOMIAL OF R(S),
      *   25H - ASCENDING POWERS OF S  )
 1013 FORMAT (1H0,5X,30HDENOMINATOR POLYNOMIAL OF R(S),
      *   25H - ASCENDING POWERS OF S  )
 1014 FORMAT (1H0,45(1H*))
 1015 FORMAT (1H0,5X,7HGAIN =  ,1PE14.7)
 1017 FORMAT(1H0,5X,35HCHARACTERISTIC POLYNOMIAL OF SYSTEM,
      *   25H - ASCENDING POWERS OF S  )
 1018 FORMAT (1H0,5X,35HTHE TIME RESPONSE OF THE STATE X(T))
 1019 FORMAT (1H0,5X,36HTHE TIME RESPONSE OF THE OUTPUT Y(T))
 1020 FORMAT (1H0,  5X,18HTHE COEFFICIENT OF    ,
      *   5H EXP(,1PE13.6,7H)T*COS(,1PE12.6,2H)T)
 1021 FORMAT (1H0,  5X,18HTHE COEFFICIENT OF    ,
      *   5H EXP(,1PE13.6,7H)T*SIN(,1PE12.6,2H)T)
 1022 FORMAT (1H0,  5X,18HTHE COEFFICIENT OF    ,
      *   5H EXP(,1PE13.6,2H)T)
 1023 FORMAT (A1,I2)
 1024 FORMAT (1H0,5X,8HRGAIN =  ,1PE14.7)
 1025 FORMAT (/6X,19HNUMERATOR ROOTS ARE/8X,9HREAL PART,
      *   10HIMAG. PART)
```

```
1027 FORMAT (/6X,21HDENOMINATCR ROCTS ARE/8X,9HREAL PART,
   *    8X,10HIMAG. PART)
 200 READ (5,1000,END=300) (NAME(I),I=1,5),NSYS
     DO 210 I=1,10
     B(I)=0.0
     FDBG(I)=0.0
     RNUM(I)=0.0
     RDEN(I)=0.0
     D(I)=0.0
     C(I)=0.0
     COEFF(I)=0.0
     EIGR(I)=0.0
     EIGI(I)=0.0
     XO(I)=0.0
     XPM(I)=0.0
     RTR(I)=C.0
     RTI(I)=0.0
     CO(I)=0.0
     DO 210 J=1,10
     A(I,J)=0.0
     PSI(1,J)=0.0
     DO 210 K=1,10
 210 CHI(I,J,K)=0.0
     PRINT 1006, (NAME(I),I=1,5)
     PRINT 1014
     PRINT 1007
     DO 100 I=1,NSYS
     READ 1001, (A(I,J),J=1,NSYS)
     PRINT 1002, (A(I,J),J=1,NSYS)
 100 CONTINUE
     READ 1001,(B(I),I=1,NSYS)
     PRINT 1008
     PRINT 1002,(B(I),I=1,NSYS)
     READ 1001, (C(I),I=1,NSYS)
     PRINT 1009
     PRINT 1002, (C(I),I=1,NSYS)
     READ 1001,(FDBG(I),I=1,NSYS)
     PRINT 1010
     PRINT 1002, (FDBG(I),I=1,NSYS)
     READ 1001, GAIN
     PRINT 1015, GAIN
     PRINT 1014
     READ 1001,(XO(I),I=1,NSYS)
     PRINT 1011
     PRINT 1002, (XO(I),I=1,NSYS)
     READ 1001, RGAIN
     PRINT 1024, RGAIN
     KOUNT=0
   1 KOUNT=KOUNT+1
     READ 1023,KEY,N
     NP=N+1
     IF(N.EQ.0) GO TO 5
     IF (KEY.EQ.IPP) GO TO 5
     I=0
   2 I=I+1
     READ 1001,RR,RI
     RTR(I)=-RR
     RTI(I)=RI
     IF(RI) 3,4,3
   3 I=I+1
     RTR(I)=-RR
     RTI(I)=-RI
```

```
   4 IF(I.LT.N) GO TO 2
     CALL SEMBL (N,RTR,RTI,CO)
     GO TO 7
   5 READ 1001,(CO(I),I=1,NP)
     CO(NP)=1.0
     IF(N) 7,7,6
   6 CALL PROOT(N,CO,RTR,RTI,+100)
   7 GO TO (80,90), KOUNT
  80 DO 81 I=1,NP
  81 RNUM(I)=CO(I)*RGAIN
     PRINT 1012
     PRINT 1002, (CO(I),I=1,NP)
     IF(N.LE.0) GO TO 1
     PRINT 1025
     DO 82 I=1,N
  82 PRINT 1002, RTR(I),RTI(I)
     GO TO 1
  90 DO 91 I=1,NP
  91 RDEN(I)=CO(I)
     LD=N
     LDR=NP
     PRINT 1013
     PRINT 1002,(CO(I),I=1,NP)
     PRINT 1027
     DO 92 I=1,N
  92 PRINT 1002, RTR(I),RTI(I)
     PRINT 1014
     N = NSYS+LD
     DO 8 I=1,N
     DO 8 J =1,N
     PSI(I,J)=0.0
   8 CONTINUE
     DO 9 I=1,NSYS
     DO 9 J=1,NSYS
     PSI(I,J) = A(I,J)-B(I)*FDBG(J)*GAIN
   9 CONTINUE
     DO 10 I=1,NSYS
     DO 10 J=1,LD
     PSI(I,NSYS+J)=B(I)*RNUM(J)*GAIN
  10 CONTINUE
     NP1 = NSYS+1
     NM1 = N-1
     DO 11 I=NP1,NM1
     PSI(I,I+1)=1.
  11 CONTINUE
     DO 12 I=1,LD
     PSI(N,NSYS+I)= -RDEN(I)
  12 CONTINUE
     CALL CHREGA(PSI,NSYS,COEFF)
     CALL PROOT(NSYS,COEFF,EIGR,EIGI,+1)
     DO 95 I=NP1,N
     J=I-NSYS
     EIGR(I)=RTR(J)
  95 EIGI(I)=RTI(J)
     CALL STMST(N,PSI,EIGR,EIGI,1)
     DO 13 I =1,NSYS
     XPM(I)=XO(I)
  13 CONTINUE
     DO 14 I =NP1,N
     XPM(I) = 0.0
  14 CONTINUE
     XPM(N) = 1.
```

```
      DO 15 I=1,N
      DO 15 J=1,NSYS
      PSI(J,I)=0.0
      DO 15 K=1,N
      PSI(J,I) = PSI(J,I)+CHI(I,J,K)*XPM(K)
   15 CONTINUE
      I = 1
      PRINT 1018
  150 IF(EIGI(I)) 16,17,16
   16 IS=I
      I=I+1
      PRINT 1003,EIGR(I),EIGI(I)
      PRINT 1002,(PSI(K,IS),K=1,NSYS)
      PRINT 1004, EIGR(I),EIGI(I)
      PRINT 1002, (PSI(K,I),K=1,NSYS)
      GO TO 18
   17 PRINT 1005, EIGR(I)
      PRINT 1002, (PSI(K,I),K=1,NSYS)
   18 IF(I-N) 19,20,20
   19 I=I+1
      GO TO 150
   20 DO 21 I=1,N
      D(I)=0.0
      DO 21 J=1,NSYS
      D(I)=D(I)+PSI(J,I)*C(J)
   21 CONTINUE
      PRINT 1014
      PRINT 1019
      I=1
   22 IF(EIGI(I)) 23,24,23
   23 IS=I
      I=I+1
      PRINT 1020,        EIGR(I),EIGI(I)
      PRINT 1002, D(IS)
      PRINT 1021,        EIGR(I),EIGI(I)
       PRINT 1002, D(I)
      GO TO 25
   24 PRINT 1022,        EIGR(I)
       PRINT 1002, D(I)
   25 IF(I-N) 26,27,27
   26 I=I+1
      GOTO 22
   27 PRINT 1014
      GO TO 200
  300 STOP
      END
```

2.4 GRAPHICAL TIME RESPONSE (GTRESP)

The graphical time response problem is used to determine and graphically display the time response of the closed-loop system

$$\dot{x}(t) = \underline{A}\underline{x}(t) + \underline{b}u(t)$$

$$u(t) = K[r(t) - \underline{k}^T\underline{x}(t)]$$

$$y(t) = \underline{c}^T\underline{x}(t)$$

due to initial conditions $\underline{x}(0)$ and an arbitrary reference input $r(t)$.

2.4-1 *THEORY*   The basic goal of the GTRESP program is quite similar to the RTRESP program of the preceding section. Here however we wish to determine the time response for arbitrary input functions which may not have rational Laplace transforms. In addition, we wish to obtain a graphical display of the response and do not care to have a closed-loop expression for the response. Since it may be desirable to have a closed-form expression and a graphical response, the GTRESP and RTRESP programs have been purposely based primarily on subroutines in order to allow maximum flexibility.

The GTRESP program is based on the use of a fourth-order Runga-Kutta numerical integration algorithm in the form of subroutines CALCU, RUNGE, TRESP and YDOT (see Appendix A) and a graphical plot routine Y8VSX (see Sec. 4.4). The input function $r(t)$ is defined by inserting coding as indicated in the subroutine CALCU. Although this graphical time response program has been written for linear systems, it may be easily modified to include nonlinear time-varying systems by appropriate modification of the subroutine YDOT.

2.4-2 *INPUT FORMAT*   The input for the GTRESP program is quite similar to that of the RTRESP program. The first part of the input contains the description of the closed-loop system while the second part contains the initial condition x(0) and the parameters for the time interval and the plot routine. The input format is summarized in Table 2.4-1.

The problem identification, system order N and $\underline{A}$, $\underline{b}$, $\underline{c}$, $\underline{k}$, K and $\underline{x}(0)$ are entered exactly as in the RTRESP on the first N+6 cards. The next card contains the time interval of interest TZERO to TF the time step to be used DT and the number of time steps between output. The final card in the input indicates which variables are to be plotted. Any eight or less of the following variables may be plotted: $y(t)$, $r(t)$, $u(t)$, $e(t) = r(t) - y(t)$, and $x_i(t)$, $i = 1, 2, \ldots, N$ by placing the symbols Y, R, U, E, I = 1, 2, ... , N in the two-column fields on the last card. If the symbol has only one character it must be right jus justified and a blank placed to the left. A blank card must be used as the last card if no variables are to be plotted.

2.4-3 *OUTPUT FORMAT*   The basic output of the GTRESP program consists of the problem identification, $\underline{A}$, $\underline{b}$, $\underline{c}$, $\underline{k}$, K and $\underline{x}(0)$. In addition the times TZERO and TF as well as DT and FREQ are printed for reference. The time response of the variables $y(t)$, $r(t)$ and $x_i(t)$, $i = 1, 2, \ldots, N$ is then printed in tabular form. Finally the graphical time response of the variables listed on the last input card is presented.

2.4-4 *EXAMPLE*   We wish to determine and plot the time response of the error $e(t) = r(t) - y(t)$, the input $r(t)$ and the state variable $x_2(t)$ for the system

$$\dot{x}(t) = \begin{bmatrix} 0.0 & 1.0 \\ -1.0 & -1.0 \end{bmatrix} \underline{x}(t) + \begin{bmatrix} 0.0 \\ 1.0 \end{bmatrix} u(t)$$

22

Table 2.4-1

Input Format for GTRESP Program

Card Number	Column Number	Description	Format
1	1-20	Problem identification	5A4,I2
	21-22	N = system order, $N \leq 10$	
2	1-10	a_{11} = Plant matrix	8F10.3
	11-20	a_{12}	
	etc.	...	
3	1-10	a_{21}	8F10.3
	11-20	a_{22}	
	etc.	...	
n+2	1-10	b_1 = control vector	8F10.3
	11-20	b_2	
	etc.	...	
n+3	1-10	c_1 = output vector	8F10.3
	11-20	c_2	
	etc.	...	
n+4	1-10	k_1 = feedback coefficients	8F10.3
	11-20	k_2	
	etc.	...	
n+5	1-10	K = controller gain	8F10.3
n+6	1-10	$x_1(0)$ = initial condition	8F10.3
	11-20	$x_2(0)$	
	etc.	...	
n+7	1-10	TZERO = initial time	8F10.3
	11-20	TF = final time	
	21-30	DT = time step	
	31-40	FREQ = frequency of output [*1]	
n+8	1-2	Any eight of the following symbols may be placed in these locations to indicate which variables are to be plotted: Y, R, U, E, 1, 2, 3, 4, 5, 6, 7, 8, 9, 10. If the symbol contains only one character, it must be right justified with a blank proceding it.	8A2
	3-4		
	5-6		
	7-8		
	9-10		
	11-12		
	13-14		
	15-16		

[1] $(TF-TZERO)/[(DT)(FREQ)] \leq 100$

$$u(t) = 5.0\{r(t) - [1.0 \quad 0.5]\underset{\sim}{x}(t)\}$$

$$y(t) = [1.0 \quad 1.0]\underset{\sim}{x}(t)$$

The time interval of interest is $0 \le t \le 4$. We let the step size be DT = 0.01 and print the output every ten steps so FREQ = 10. We let the initial conditions conditons be $\underset{\sim}{x}(0) = \underset{\sim}{0}$ and the input function is

$$r(t) = 5.0 \qquad \text{if} \quad 0 \le t \le 1$$
$$ = 0 \qquad \text{otherwise}$$

The input card deck for this example is given in Table 2.4-2. In order to define the input we add the coding to CALCU

```
1000    IF(T.GT.1.0) GO TO 1001
        R = 5.0
        GO TO 1002
1001    R = 0.0
1002    CONTINUE
```

Table 2.4-2

Input Data Deck for GTRESP Example

Card No.	Column 5	10	15	20	25	30
1	GTRESP	EXAMPLE		2		1.0
2	0.0	-1.0				0.01
3	0.0	-1.0				
4	0.0	1.0				
5	0.0	-1.0				
6	5.0	0.5				
7	0.0	0.0				
8	0.0	4.0				
9	0.0					
10	ER	2				

The program output is shown in Table 2.4-3. The data points on the graphical response have been joined to facilitate viewing of the results.

Table 2.4-3

Program Output for GTRESP Example

GRAPHICAL TIME RESPONSE
PROBLEM IDENTIFICATION – GTRESP EXAMPLE

****

      THE A MATRIX

0.0   1.00000000E 00
-1.00000000E 00   -1.00000000E 00

      THE B MATRIX

0.0   1.00000000E 00

      THE C MATRIX

1.00000000E 00   1.00000000E 00

      FEEDBACK COEFF.

1.00000000E 00   5.00000000E-01

      GAIN =   5.00000000E 00

      INITIAL CONDITIONS

0.0   0.0

      TZERO =   0.0   TF =   4.000000
      DT =   0.010000   FREQ =   10

****

T	Y(T)	U(T)	X1(T)	X2(T)
0.0	0.0	2.500000E 01	0.0	0.0
9.999990E-02	2.199443E 00	1.922372E 01	1.110644E-01	2.088379E
1.999998E-01	3.848458E 00	1.439480E 01	3.936214E-01	3.454837E
2.999997E-01	5.026528E 00	1.047669E 01	7.827941E-01	4.243734E
3.999996E-01	5.813385E 00	7.397809E 00	1.227492E 00	4.585893E
4.999995E-01	6.284925E 00	5.065398E 00	1.688916E 00	4.596008E
5.999994E-01	6.510508E 00	3.376522E 00	2.138885E 00	4.371623E
6.999993E-01	6.551410E 00	2.226191E 00	2.558115E 00	3.993295E
7.999992E-01	6.460109E 00	1.513309E 00	2.934568E 00	3.525540E
8.999991E-01	6.280244E 00	1.144614E 00	3.261911E 00	3.018332E
9.999990E-01	6.047029E 00	1.037025E 00	3.538161E 00	2.508868E
1.099982E 00	3.620378E 00	-1.819321E 01	3.656905E 00	-3.652740E
1.199965E 00	1.698620E 00	-1.313748E 01	3.556374E 00	-1.857754E
1.299949E 00	2.588177E-01	-8.913007E 00	3.306385E 00	-3.047567E

3.499579E 00	8.149618E-02	-1.150310E-02	-7.689488E-02	1.583911E
3.599563E 00	8.478630E-02	-5.793137E-02	-6.161377E-02	1.464001E
3.699546E 00	8.283365E-02	-8.772963E-02	-4.774184E-02	1.305755E
3.799529E 00	7.728511E-02	-1.043089E-01	-3.556158E-02	1.128467E
3.899512E 00	6.948203E-02	-1.107377E-01	-2.518698E-02	9.466904E
3.999496E 00	6.048328E-02	-1.096897E-01	-1.660740E-02	7.709068E

25

Table 2.4-3 (cont.)

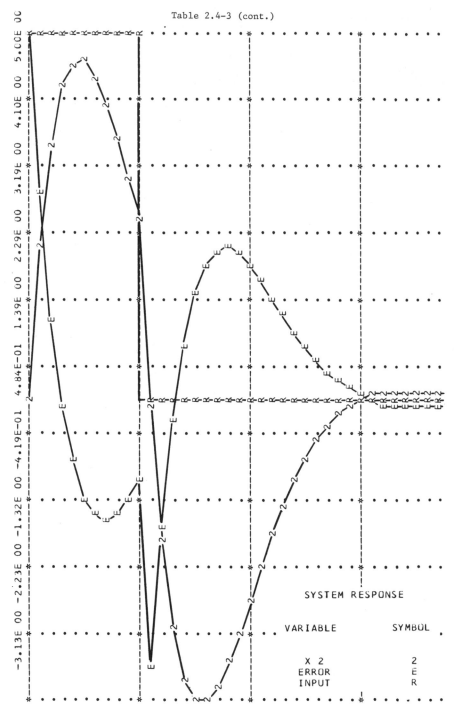

SYSTEM RESPONSE

VARIABLE	SYMBOL
X 2	2
ERROR	E
INPUT	R

2.4-5 *SUBPROGRAMS USED* The following subprograms are used by this program:

    (1)   CALCU
    (2)   RUNGE
    (3)   TRESP
    (4)   YDOT
    (5)   Y8VSX

See Appendix A and Sec. 4.4 for a description of each of the subprograms.

2.4-6 *PROGRAM LISTING*

```
C     GRAPHICAL TIME RESPONSE (GTRESP)
C     SUBPROGRAMS USED: CALCU, RUNGE, TRESP, YDOT
      INTEGER CHAR(15)
      COMMON IPLOT,IVAR(10)
      DIMENSION A(10,10),C(10),B(10), AK(10),X(10),NAME(5)
      DATA CHAR(1),CHAR(2),CHAR(3),CHAR(4),CHAR(5),
     *CHAR(6),CHAR(7),CHAR(8),CHAR(9),CHAR(10),CHAR(11),
     *CHAR(12),CHAR(13),CHAR(14),CHAR(15)/2H 1,2H 2,2H 3,2H 4,
     *2H 5,2H 6,2H 7,2H 8,2H 9,2H10,2H E,2H U,2H Y,2H R,2H  /
    3 FORMAT (8F10.3)
 1000 FORMAT (1H0, 10X, 8HTZERO = , F10.6,10X, 5HTF = ,F10.6/
     * 11X,5HDT = ,F10.6,13X,7HFREQ = I5)
 1001 FORMAT (1H0,10X,13H THE A MATRIX   /)
 1002 FORMAT (6(1PE20.8))
 1003 FORMAT (1H0,10X,19H INITIAL CONDITIONS    /)
 1004 FORMAT (1H0,10X,13H THE B MATRIX   /)
 1005 FORMAT (1H0,10X,16H FEEDBACK COEFF.     /)
 1006 FORMAT (1H0,10X,8H GAIN = , 1PE20.8  )
 1007 FORMAT (1H0,10X,13H THE C MATRIX    /)
 1008 FORMAT (8A2)
 1009 FORMAT( 5X,25HPROBLEM IDENTIFICATION - ,5A4)
 1010 FORMAT(1H1,4X,23HGRAPHICAL TIME RESPONSE)
 1011 FORMAT(/5X,45(1H*))
   10 READ(5,1,END=20)(NAME(I),I=1,5),N
    1 FORMAT(5A4,I2)
      DO 60 I=1,8
   60 IVAR(I)=CHAR(15)
      PRINT 1010
      PRINT 1009,(NAME(I),I=1,5)
      PRINT 1011
      PRINT 1001
      DO 2 I = 1, N
      READ 3, (A(I,J), J=1,N)
      PRINT 1002, (A(I,J), J=1,N)
    2 CONTINUE
      READ 3, (B(I),I=1,N)
      PRINT 1004
      PRINT 1002, (B(I),I=1,N)
      READ 3,(C(I),I=1,N)
      PRINT 1007
      PRINT 1002, (C(I),I=1,N)
      READ 3, (AK(I), I=1,N)
      PRINT 1005
      PRINT 1002, (AK(I),I=1,N)
      READ 3, GAIN
      PRINT 1006, GAIN
      READ 3, (X(I),I=1,N)
      PRINT 1003
      PRINT 1002, (X(I),I=1,N)
      READ 3, TZERO, TF,DT,FREQ
```

```
      IFQ=FREQ
      PRINT 1000, TZERO, TF ,DT,IFQ
      PRINT 1011
      READ 1008, (IVAR(I),I=1,8)
      DO 40 I=1,8
      DO 30 J=1,15
      IF (IVAR(I)-CHAR(J)) 30,25,30
   25 IVAR(I)=J

      GO TO 40
   30 CONTINUE
   40 CONTINUE
      MIN=1
      MAX=8
      M=8
  419 DO 42 I=MIN,MAX
      IF(IVAR(I).NE.15) GO TO 42
      M=MAX-1
      IF(I.GT.M) GO TO 42
      DO 43 J=I,M
   43 IVAR(J)=IVAR(J+1)
      GO TO 431
   42 CONTINUE
      GO TO 432
  431 MIN=I
      MAX=M
      GO TO 419
  432 IPLOT = M
      IF(IPLOT.LT.2) GO TO 50
      LIM=IPLOT-1
      DO 44 I=1,LIM
      MIN=I+1
      DO 44 J=MIN,IPLOT
      IF(IVAR(I)-IVAR(J)) 44,44,45
   45 IHOLD=IVAR(I)
      IVAR(I)=IVAR(J)
      IVAR(J)=IHOLD
   44 CONTINUE
   50 CALL TRESP(A,X,B,AK,TZERO,TF,DT,IFQ,N,GAIN,C)
      GO TO 10
   20 STOP
      END
```

2.5 SENSITIVITY ANALYSIS (SENSIT)

The SENSIT program is used to study the variation of the closed-loop poles of the linear feedback system

$$\dot{\underline{x}}(t) = \underline{A}\underline{x}(t) + \underline{b}u(t)$$

$$u(t) = K[r(t) - \underline{k}^T\underline{x}(t)]$$

as an element of $\underline{A}$, $\underline{b}$, $\underline{k}$ or the gain K varies.

2.5-1 *THEORY* We could obtain the sensitivity analysis of the system by forming $\underline{A}_k = \underline{A} - K\underline{b}\underline{k}^T$ for the various parameter values and then using the subroutines CHREQ and PROOT to obtain and factor the closed-loop characteristic polynomial. However by considering the parameter variations more carefully, it is possible to avoid much of the computations involved in the determination of the characteristic polynomial $D_k(s)$. The subroutine PROOT is still used to factor $D_k(s)$ for each parameter value and hence the computer time can become large for high order systems.

As noted above, this program permits a variation in one element of $\underline{A}$, $\underline{b}$, $\underline{k}$ or K. Four separate sections of coding are used to handle each of the possible variations. Let us consider each of these sections.

Variation of K: The program calculates the two polynomials

$$D_p(s) = \det(s\underline{I} - \underline{A})$$

and

$$N_h(s) = \underline{k}^T \text{adj}(s\underline{I} - \underline{A})\underline{b}$$

and then forms $D_k(s)$ as a function of K as

$$D_k(s) = D_p(s) + KN_h(s)$$

In this fashion $D_p(s)$ and $N_h(s)$ do not need to be recalculated as K varies. The polynomial $D_p(s)$ is obtained directly from CHREQ and the adj$(s\underline{I} - \underline{A})$ is simultaneously obtained as

$$\text{adj}(s\underline{I} - \underline{A}) = \underline{\Psi}_1 + \underline{\Psi}_2 s + \ldots + \underline{\Psi}_{n-2}s^{n-1}$$

The coefficients of $N_h(s)$ are then given as

$$N_h(s) = \underline{k}^T\underline{\Psi}_1\underline{b} + \underline{k}^T\underline{\Psi}_2\underline{b}s + \ldots + \underline{k}^T\underline{\Psi}_{n-2}\underline{b}s^{n-1}$$

Variation of k_i: In this case, the program again calculates $\underline{\Phi}(s)$ in the form

$$\underline{\Phi}(s) = \frac{\text{adj}(s\underline{I} - \underline{A})}{\det(s\underline{I} - \underline{A})}$$

by the use of CHREQ. From adj$(s\underline{I} - \underline{A})$, the vector polynomial $\underline{w}(s)$ is computed as

$$\underline{w}(s) = \text{adj}(s\underline{I} - \underline{A})\underline{b}$$

Then k_i is set to zero and the polynomial $P_1(s)$ is formed

$$P_1(s) = \underline{k}^T\underline{w}(s)$$

The characteristic polynomial $D_k(s)$ is now

$$D_k(s) = D_p(s) + K[P_1(s) + k_iw_i(s)]$$

and only a simple scalar computation of k_i times $w_i(s)$ needs to be redone for each value of k_i.

Variation of b_i: The variation of b_i is handled in much the same manner as the variation of k_i. After $\Phi(s)$ is computed as above, the vector polynomial $\underset{\sim}{q}(s)$ is computed as

$$\underset{\sim}{q}^T(s) = \underset{\sim}{k}^T \text{ adj}(s\underset{\sim}{I} - \underset{\sim}{A})$$

Then with $b_i = 0$, the scalar polynomial $R_1(s)$ is formed

$$R_1(s) = \underset{\sim}{q}^T(s)\underset{\sim}{b}$$

and the closed-loop characteristic polynomial is now

$$D_k(s) = D_p(s) + K[R_1(s) + q_i(s)b_i]$$

Variation of a_{ij}: In the case of a variation of a_{ij}, there is little which can be done to simplify the computations of $D_k(s)$. For each value of a_{ij}, the resolvent matrix $\Phi(s)$ is formed and then $D_k(s)$ is determined as

$$D_k(s) = D_p(s) + K\underset{\sim}{k}^T \text{adj}(s\underset{\sim}{I} - \underset{\sim}{A})\underset{\sim}{b}$$

The variation of a_{ij} uses the most computer time of the four possible cases which the program can treat since the resolvent matrix must be recomputed for each parameter value.

Linear interpolation is used to compute the parameter values between the extreme values which are read into the program. There are certain cases[1] where a logarithmic, geometric or a combination of several types of interpolation may be more useful. Because of the generality of this program, it did not appear to be possible to select a specific type of interpolation which would be universally best. Hence the simplest form was used.

This program may be used to study the variation of the closed-loop poles for several different parameters in the same system. After completing the root locus plot for a given parameter variation, the program resets this parameter to its initial value and returns to read a card identifying the next parameter to be varied. A blank card is used to return the program to start a new problem.

2.5-2 *INPUT FORMAT* The first N+4 input cards consists of the problem identifications, system order, $\underset{\sim}{A}$, $\underset{\sim}{b}$, $\underset{\sim}{k}$ and K information in the usual format. The next input cards define parameter variations which are to be studied and the final card is blank to signal the termination of the problem input for this system. The complete input format is summarized in Table 2.5-1.

On the cards which indicate the parameter study to be made, the first column is used to indicate whether an element of $\underset{\sim}{A}$, $\underset{\sim}{b}$, $\underset{\sim}{k}$ or K is to be varied by the symbol A, B, K or G respectively. The next two two-column fields are used to indicate the exact element to be varied. If the parameter is an element of $\underset{\sim}{A}$, a_{ij}, then i and j are placed in these two fields. If an element of either $\underset{\sim}{b}$ or $\underset{\sim}{k}$ is to be varied, then the index is placed in column 2-3 and 4-5 are left blank. If K is to be varied, then columns 2-5 are left blank.

As indicated previously any number of parameter studies may be made for the same system by using a number of different cards of the form described above. The final card in the input data deck for a given problem must be blank to indicate the end of data for that problem.

---

[1] See L. P. Huelsman: <u>Digital Computations in Basic Circuit Theory</u>, McGraw-Hill Book Company, Inc., 1968.

Table 2.5-1

Input Format for SENSIT Program

Card Number	Column Number	Description	Format
1	1-20	Problem identification	5A4,I2
	21-22	N = system order, N $\leq$ 10	
2	1-10	a_{11} = Plant matrix	8F10.3
	11-20	a_{12}	
	etc.	...	
3	1-10	a_{21}	8F10.3
	11-20	a_{22}	
	etc.	...	
N+2	1-10	b_1 = control vector	8F10.3
	11-20	b_2	
	etc.	...	
N+3	1-10	k_1 = feedback coefficients	8F10.3
	11-20	k_2	
	etc.	...	
N+4	1-10	K = controller gain	8F10.3
N+5	1	IVAR = element to be varied.	A1,2I2
		= A for plant matrix A	I5,2F10.3
		= B for control vector b	
		= K for feedback coefficient k	
		= G for controller gain K	
	2-3	I1 = i for a_{ij}, b_i, k_i	
	4-5	I2 = j for a_{ij}	
	6-10	NUMG = number of parameter values to be used	
	11-20	GMIN = minimum value of parameter	
	21-30	GMAX = maximum value of parameter.	
Final	—	BLANK CARD	

2.5-3 *OUTPUT FORMAT*   For reference, the problem identification as well as A, b, k and K are printed at the beginning of the output. Then the parameter to be varied and its minimum and maximum values are given. Following this the closed-loop poles associated with each parameter value are listed. Finally a root locus plot of the closed-loop pole location on the complex s plane is provided for graphical representation of the sensitivity study.

2.5-4 *EXAMPLE*   Let us consider the third-order linear system given by

$$\dot{\underset{\sim}{x}}(t) = \begin{bmatrix} 0 & 1 & 0 \\ 0 & -2 & 2 \\ 0 & 0 & -8 \end{bmatrix} \underset{\sim}{x}(t) + \begin{bmatrix} 0 \\ 0 \\ 5 \end{bmatrix} u(t)$$

$$u(t) = 12\{r(t) - [1 \quad 0.283 \quad 0.15]\underset{\sim}{x}(t)\}$$

We wish to study the effect on the closed-loop poles of varying the element a_{22} from 0 to -10 in steps of 0.5. The complete input data deck for this problem is shown in Table 2.5-2. The resulting program output is given in Table 2.5-3.

Table 2.5-2

Input Data Deck for SENSIT Example

Card No.	Column No.				
	5	10	15	20	25	30
1	S E N S I T	E X A M P L E	O N E	3		
2	0 . 0	1 . 0		0 . 0		
3	0 . 0	- 2 . 0		2 . 0		
4	0 . 0	0 . 0		- 8 . 0		
5	0 . 0	0 . 0		5 . 0		
6	1 . 0	. 2 8 3 3 3		. 1 5		
7	1 2 . 0					
8	A 2 2	2 1 - 1 0 . 0		0 . 0		
9		(B L A N K C A R D)				

Only a small portion of the graphical output for the root locus plot is shown. The values of a_{22} associated with the closed-loop have been read off of the tabular output and placed on the plot for easy reference.

Table 2.5-3

Partial Program Output for SENSIT Example

SENSITIVITY ANALYSIS PROGRAM
PROBLEM IDENTIFICATION - SENSIT EXAMPLE ONE

THE A MATRIX
0.0   1.0000E 00   0.0
0.0   -2.0000E 00   2.0000E 00
0.0   0.0   -8.0000E 00

THE B MATRIX

0.0   0.0   5.0000E 00

FEEDBACK COEFFICIENTS

1.0000E 00   2.8333E-01   1.5000E-01

GAIN = 1.2000E 01

32

Table 2.5-3 (cont.)

ROOT LOCUS AS A(2, 2) VARIES BETWEEN -1.0000E 01 AND   0.0

A(2, 2) = -1.0000E 01
ROOTS ARE

REAL PART	IMAG. PART
-1.3179E 01	-3.6606E 00
-1.3179E 01	3.6606E 00
-6.4139E-01	0.0

A(2, 2) = -9.5000E 00
ROOTS ARE

REAL PART	IMAG. PART
-1.2913E 01	-3.3691E 00
-1.2913E 01	3.3691E 00
-6.7378E-01	0.0

A(2, 2) = -9.0000E 00
ROOTS ARE

REAL PART	IMAG. PART
-1.2645E 01	-3.0249E 00
-1.2645E 01	3.0249E 00
-7.0986E-01	0.0

A(2, 2) = -2.0000E 00
ROOTS ARE

REAL PART	IMAG. PART
-1.5000E 01	0.0
-2.0000E 00	-2.0000E 00
-2.0000E 00	2.0000E 00

A(2, 2) = -1.5000E 00
ROOTS ARE

REAL PART	IMAG. PART
-1.5083E 01	0.0
-1.7087E 00	-2.2442E 00
-1.7087E 00	2.2442E 00

A(2, 2) = -1.0000E 00
ROOTS ARE

REAL PART	IMAG. PART
-1.5158E 01	0.0
-1.4212E 00	-2.4284E 00
-1.4212E 00	2.4284E 00

A(2, 2) = -5.0000E-01
ROOTS ARE

REAL PART	IMAG. PART
-1.5226E 01	0.0
-1.1368E 00	-2.5669E 00
-1.1368E 00	2.5669E 00

A(2, 2) = 0.0
ROOTS ARE

REAL PART	IMAG. PART
-1.5290E 01	0.0
-8.5521E-01	-2.6678E 00
-8.5521E-01	2.6678E 00

Table 2.5-3 (cont.)

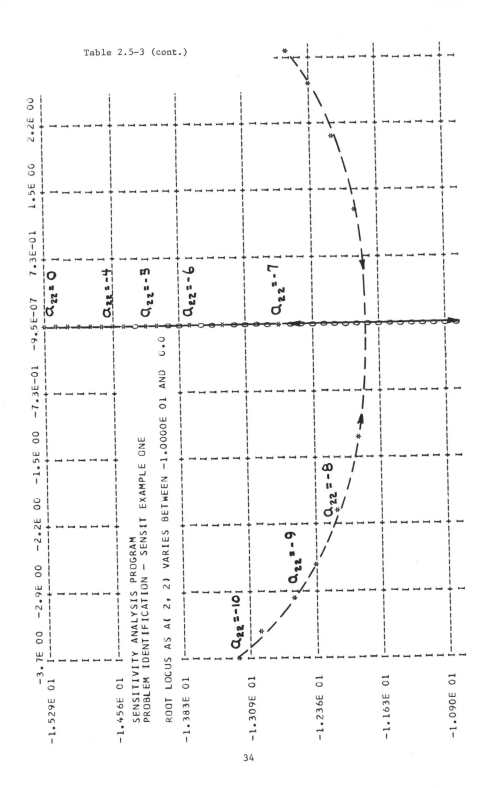

SENSITIVITY ANALYSIS PROGRAM
PROBLEM IDENTIFICATION — SENSIT EXAMPLE ONE

ROOT LOCUS AS A(2, 2) VARIES BETWEEN -1.0000E 01 AND 0.0

2.5-5 *SUBPROGRAMS USED*   The following subprograms are used by this
program

    (1) CHREQ
    (2) FORM
    (3) MPY
    (4) PROOT
    (5) SPLIT

See Appendix A and Sec. 4.3 for a description of each of these subprograms.

2.5-6 *PROGRAM LISTING*

```
C  SENSITIVITY ANALYSIS PROGRAM  (SENSIT)
C  SUBPROGRAMS USED:  CHREQ, FORM, MPY, PROOT, SPLIT.
       COMMON ZED(10,10,10)
       DIMENSION A(10,10),B(10),G(10),RR(10),RI(10),
      *ANS(11),C(11),U(999),V(999),NAME(5)
       DIMENSION D(11)
       DIMENSION W(10,10),E(11)
       DATA IAA,IKK,IBB,IGG,IBLANK/1HA,1HK,1HB,1HG,1H /
 1000 FORMAT (A1,2I2,I5,2F10.3)
 1001 FORMAT (5A4,I2)
 1002 FORMAT (8F10.3)
 1003 FORMAT (/5X,2HA(,I2,1H,,I2,4H) = ,1PE11.4,/5X,
      *9HROOTS ARE/5X,9HREAL PART,5X,10HIMAG. PART,/
      *(5X,1PE11.4,5X,1PE11.4))
 1004 FORMAT (/5X,16HROOT LOCUS AS A(,I2,1H,,I2,
      *  17H) VARIES BETWEEN ,1PE11.4,5H AND ,1PE11.4)
 1005 FORMAT(1H1,4X,28HSENSITIVITY ANALYSIS PROGRAM/5X,
      *25HPROBLEM IDENTIFICATION - ,5A4)
 1006 FORMAT (/5X,12HTHE A MATRIX)
 1007 FORMAT (8(5X,1PE11.4))
 1008 FORMAT (/5X,12HTHE B MATRIX//10(5X,1PE11.4))
 1009 FORMAT (/5X,21HFEEDBACK COEFFICIENTS//10(5X,1PE11.4))
 1010 FORMAT (/5X,7HGAIN = ,1PE11.4)
 1011 FORMAT(/5X,16HROOT LOCUS AS B(,I2,17H) VARIES BETWEEN ,
      *1PE11.4,5H AND ,1PE11.4)
 1012 FORMAT (/5X,2HB(,I2,4H) = ,1PE13.6/5X,9HROOTS ARE/
      *5X,9HREAL PART,5X,10HIMAG. PART/(5X,1PE11.4,5X,1PE11.4))
 1013 FORMAT(/5X,16HROOT LOCUS AS K(,I2,17H) VARIES BETWEEN ,
      *  1PE11.4,5H AND ,1PE11.4)
 1014 FORMAT (/5X,2HK(,I2,4H) = ,1PE13.6,/5X,9HROOTS ARE/
      *5X,9HREAL PART,5X,10HIMAG. PART/(5X,1PE11.4,2X,1PE11.4))
 1015 FORMAT (/5X,38HROOT LOCUS AS THE GAIN VARIES BETWEEN ,
      *1PE11.4,5H AND ,1PE11.4)
 1016 FORMAT(/5X,7HGAIN = ,1PE13.6/5X,9HROOTS ARE/5X,
      *  9HREAL PART,5X,10HIMAG. PART/(5X,1PE11.4,5X,1PE11.4))
       NPRT=100
       ICK=100
  100 READ (5,1001,END=2000) (NAME(I),I=1,5),N
       DO 10 I=1,10
       B(I)=0.0
       G(I)=0.0
       RR(I)=0.0
       RI(I)=0.0
       DO 10 J=1,10
       A(I,J)=0.0
       W(I,J)=0.0
       DO 10 K=1,10
   10 ZED(I,J,K)=0.0
       DO 20 J=1,11
       ANS(J)=0.0
       C(J)=0.0
```

```
      D(J)=0.0
   20 E(J)=0.0
      DO 30 J=1,999
      U(J)=0.0
   30 V(J)=0.0
      DO 101 I=1,N
  101 READ 1002, (A(I,J),J=1,N)
      READ 1002,(B(I),I=1,N)
      READ 1002,(G(I),I=1,N)
      READ 1002, GAIN
  102 READ 1000,IVAR,I1,I2,NPTS,GMIN,GMAX
      IF(NPTS.GT.1) GO TO 103
      NPTS=2
  103 XNUMG=NPTS
      DELTA=(GMAX-GMIN)/(XNUMG-1.0)
      IF(IVAR.EQ.IBLANK) GO TO 100
      PRINT 1005, NAME
      PRINT 1006
      DO 104 I=1,N
  104 PRINT 1007, (A(I,J),J=1,N)
      PRINT 1008,(B(I),I=1,N)
      PRINT 1009, (G(I),I=1,N)
      PRINT 1010, GAIN
      IF(IVAR.EQ.IBB) GO TO 300
      IF (IVAR.EQ.IKK) GO TO 400
      IF (IVAR.EQ. IGG) GO TO 500
  200 HOLD = A(I1,I2)
      ITER=0
      PRINT 1004,I1,I2,GMIN,GMAX
      A(I1,I2)=GMIN
  201 CALL CHREQ(A,N,C,NPRT)
      CALL MPY(B,G,N,ANS,NS)
      CALL FORM(GAIN,ANS,NS,C,N,D,ID)
      CALL PROOT (ID,D   ,RR,RI,ICK)
      PRINT 1003, I1,I2,A(I1,I2),(RR(I),RI(I),I=1,ID)
      M=ID*ITER
      ITER=ITER+1
      DO 202 I=1,ID
      U(M+I)=RR(I)
  202 V(M+I)=RI(I)
      IF(A(I1,I2).GE.GMAX) GO TO 203
      A(I1,I2)=A(I1,I2)+DELTA
      GO TO 201
  203 NPTS=ID*ITER
      A(I1,I2)=HOLD
      PRINT 1005, NAME
      PRINT 1004,I1,I2,GMIN,GMAX
      CALL SPLIT (U,V,NPTS)
      GO TO 102
  300 HOLD=B(I1)
      ITER=0
      PRINT 1011,I1,GMIN,GMAX
      B(I1)=0.0
      CALL CHREQ(A,N,C,NPRT)
      NM=N-1
      DO 306 I=1,N
      DO 306 J=1,N
      W(I,J)=0.0
      DO 306 K=1,N
  306 W(I,J)=W(I,J)+ZED(I,K,J)*G(K)
      DO 307 I=1,N
      ANS(I)=0.0
```

```
      DO 307 J=1,N
307 ANS(I)=ANS(I)+W(I,J)*B(J)
    VAR=GMIN
309 DO 308 I=1,N
308 D(I)=ANS(I)+VAR*W(I,I1)
    CALL FORM(GAIN,D,NM,C,N,E,ID)
    CALL PROOT(ID,E,RR,RI,ICK)
    PRINT 1012,I1,VAR  ,(RR(I),RI(I),I=1,ID)
    M=ID*ITER
    ITER=ITER+1
    DO 302 I=1,ID
    U(M+I)=RR(I)
302 V(M+I)=RI(I)
    IF(VAR.GE.GMAX) GO TO 303
    VAR=VAR+DELTA
    GO TO 309
303 NPTS=ID*ITER
    B(I1)=HOLD
    PRINT 1005, NAME
    PRINT 1011, I1,GMIN,GMAX
    CALL SPLIT(U,V,NPTS)
    GO TO 102
400 HOLD=G(I1)
    ITER = 0
    PRINT 1013,I1,GMIN,GMAX
    G(I1)=0.0
    CALL CHREQ(A,N,C,NPRT)
    NM=N-1
    DO 405 I=1,N
    DO 405 J=1,N
    W(I,J)=0.0
    DO 405 K=1,N
405 W(I,J)=W(I,J)+ZED(I,J,K)*B(K)
    DO 406 I=1,N
    ANS(I)=0.0
    DO 406 J=1,N
406 ANS(I)=ANS(I)+W(I,J)*G(J)
    VAR=GMIN
407 DO 408 I=1,N
408 D(I)=ANS(I)+VAR*W(I,I1)
     CALL FORM(GAIN,D,NM,C,N,E,NE)
    CALL PROOT(NE,E,RR,RI,ICK)
    ID=NE
    NS=NE
    PRINT 1014, I1,VAR  ,(RR(I),RI(I),I=1,ID)
    M=ID*ITER
    ITER=ITER+1
    DO 402 I=1,ID
    U(M+I)=RR(I)
402 V(M+I)=RI(I)
    IF(VAR.GE.GMAX) GO TO 403
    VAR=VAR+DELTA
    GO TO 407
403 NPTS=ID*ITER
    G(I1)=HOLD
    PRINT 1005, NAME
    PRINT 1013,I1,GMIN,GMAX
    CALL SPLIT(U,V,NPTS)
    GO TO 102
500 HOLD=GAIN
    ITER=0
```

```
       PRINT 1015,GMIN,GMAX
       GAIN=GMIN
       CALL CHREQ(A,N,C,NPRT)
       CALL MPY(B,G,N,ANS,NS)
  501  CALL FORM(GAIN,ANS,NS,C,N,D,ID)
       CALL PROOT(ID,D   ,RR,RI,ICK)
       PRINT 1016,GAIN,(RR(I),RI(I),I=1,ID)
       M=ID*ITER
       ITER=ITER+1
       DO 502 I=1,ID
       U(M+I)=RR(I)
  502  V(M+I)=RI(I)
       IF(GAIN.GE.GMAX) GO TO 503
       GAIN=GAIN+DELTA
       GO TO 501
  503  NPTS=ID*ITER
       GAIN=HOLD
        PRINT 1005, NAME
       PRINT 1015, GMIN,GMAX
       CALL SPLIT(U,V,NPTS)
       GO TO 102
 2000  STOP
       END
```

2.6 STATE VARIABLE FEEDBACK (STVARFDBK)

From the plant matrix $\underset{\sim}{A}$, the control vector $\underset{\sim}{b}$, and the output vector, $\underset{\sim}{c}$, the program computes the plant transfer function $G_p(s) = \underset{\sim}{c}^T\Phi(s)\underset{\sim}{b}$. The program can determine internal transfer functions[1] of the form $x_i(s)/u(s)$ by simply defining a fictitious $\underset{\sim}{c}$ vector. If, for example, the transfer function $x_3(s)/u(s)$ is desired then $\underset{\sim}{c}$ is selected with $c_3 = 1$ and $c_j = 0$ for $j \neq 3$. Any number of these internal transfer functions may be determined at the same time. However, the $\underset{\sim}{c}$ vector for the actual output y must always be used last if closed-loop computations are to be made. The denominator polynomial of all transfer functions of the form $x_i(s)/u(s)$ is always the plant characteristic polynomial $D_p(s)$. If a transfer function of the form $x_i(s)/x_j(s)$ is needed, then one can simply divide $x_i(s)/u(s)$ by $x_j(s)/u(s)$.

If, in addition to $\underset{\sim}{A}$, $\underset{\sim}{b}$ and $\underset{\sim}{c}$, the controller gain K and feedback coefficient vector $\underset{\sim}{k}$ are supplied, then the program computes the closed-loop transfer function $y(s)/r(s)$ and the equivalent feedback transfer function $H_{eq}(s)$.

The program can also be used for design by supplying the desired closed-loop transfer function. The controller gain and feedback coefficients will be calculated along with $H_{eq}(s)$. The controller gain will be selected so that zero steady state error for step inputs results.

2.6-1 *THEORY* This program is based on the realization that if the plant is represented in phase variables, then the computational results described above may be obtained by inspection[2]. In order to illustrate this feature, consider the phase variable representation of the open-loop described by the transfer function

$$G_p(s) = \frac{KN_p(s)}{D_p(s)} = \frac{c_1 + c_2 s + \ldots + c_m s^{m-1}}{a_1 + a_2 s + \ldots + s^n} \qquad (2.6\text{-}1)$$

which takes the phase variable form [3]

$$\dot{\underset{\sim}{x}}^P(t) = \underset{\sim}{A}^P \underset{\sim}{x}^P(t) + \underset{\sim}{b}^P u(t)$$

$$y(t) = (\underset{\sim}{c}^P)^T \underset{\sim}{x}^P(t) \qquad (2.6\text{-}2)$$

where

$$\underset{\sim}{A}^P = \begin{bmatrix} 0 & 1 & 0 & \ldots & 0 \\ 0 & 0 & 1 & \ldots & 0 \\ \vdots & & & & \vdots \\ 0 & 0 & 0 & \ldots & 1 \\ -a_1 & -a_2 & -a_3 & \ldots & -a_n \end{bmatrix} \qquad (2.6\text{-}3)$$

$$\underset{\sim}{b}^P = col(0,0,\ldots 0,1) \qquad\qquad \underset{\sim}{c}^P = col(c_1,\ldots,c_m,0,\ldots,0)$$

---

[1] The use of internal transfer functions has been shown to be a valuable aid in the design of closed-loop systems. See, for example, H. S. Murray and J. L. Melsa, "State-Variable Feedback Control of Coupled Nuclear Systems," presented at the ANS National Topical Meeting on Coupled Reactor Kinetics, Texas A&M, January 22-23, 1967.

[2] J. L. Melsa, "An Algorithm for the Analysis and Design of Linear State Variable Feedback Systems," <u>Proc. of Asilomar Conference on Circuits and Systems</u>, Monterey, California, pp. 791-799, Nov. 1967.

[3] The superscript "p" is used throughout to indicate phase variable quantities.

A comparison of Eq. (2.6-1) and Eqs. (2.6-2) and (2.6-3) reveals that the plant transfer function may be determined by inspection from the phase variable representation. Internal transfer functions of the form $x_i(s)/u(s)$ may be determined by simply defining a fictitious $\underset{\sim}{c}$ matrix with $c_i = 1$ and $c_j = 0$ for $j \neq i$.

In terms of the closed-loop calculations, the task may be similarly accomplished by inspection. If the closed-loop expression for the control is given by

$$u(t) = K[r(t) - (\underset{\sim}{k}^P)^T \underset{\sim}{x}^P(t)]$$

where $\underset{\sim}{k}^P$ is the feedback coefficient vector in terms of phase variables, then the system representation becomes

$$\underset{\sim}{\dot{x}}^P(t) = \begin{bmatrix} 0 & 0 & \cdots & 0 \\ 0 & 0 & \cdots & 0 \\ \cdots\cdots\cdots\cdots\cdots\cdots\cdots\cdots\cdots\cdots\cdots\cdots\cdots\cdots \\ 0 & 0 & \cdots & 0 \\ -(a_1 + Kk_1{}^P) & -(a_2 + Kk_2{}^P) & \cdots & -(a_n + Kk_n{}^P) \end{bmatrix} \underset{\sim}{x}^P(t) + \begin{bmatrix} 0 \\ 0 \\ \cdots \\ 0 \\ 1 \end{bmatrix} r(t)$$

and

$$y(t) = K[c_1 \ \cdots \ c_m \quad 0 \ \cdots \ 0]\underset{\sim}{x}^P(t)$$

Therefore the closed-loop transfer function is given by

$$\frac{y(s)}{r(s)} = \frac{K(c_1 + c_2 s + \ldots + c_m s^{m-1})}{(a_1 + Kk_1{}^P) + (a_2 + Kk_2{}^P)s + \ldots + (a_n + Kk_n{}^P)s^{n-1} + s^n}$$

The coefficients of the denominator of $y(s)/r(s)$ may be determined from the simple equations

$$a_{ki} = a_i + Kk_i{}^P \qquad i = 1, 2, \ldots, n$$

where

$$\frac{y(s)}{r(s)} = \frac{K(c_1 + c_2 s + \ldots + c_m s^{m-1})}{a_{k1} + a_{k2}s + \ldots + a_{kn}s^{n-1} + s^n}$$

Hence, if $\underset{\sim}{k}^P$ and K are known, $y(s)/r(s)$ may be determined directly. Similarly if the a_i's and a_{ki}'s are known then K and $\underset{\sim}{k}^P$ may also be obtained directly. In order to make the solution for K and $\underset{\sim}{k}^P$ unique, steady-state position error may be set to zero so that K is given by

$$K = \frac{a_{k1}}{c_1}$$

In addition to the above advantages of the phase variable representation, it is easy to show that the equivalent feedback transfer function $H_{eq}(s)$ is also easily related to the phase variable elements by the following equation.

$$H_{eq}(s) = \frac{k_1 P + k_2 P_s + \ldots + k_n P_s^{n-1}}{c_1 + c_2 s + \ldots + c_m s^{m-1}}$$

Most plants are not naturally described in phase variables, and it is not recommended that the plant representation be limited to phase variables since this would destroy the generality of the matrix formulation. The phase variable approach may still be used by transforming the plant to phase variables and then transforming the feedback coefficients back to the original state variable representation. Kalman[1] has shown that it is possible to transform any controllable plant to phase variables by means of a nonsingular linear transformation of variables of the form

$$\underset{\sim}{x} = \underset{\sim}{P} x^P$$

or since $\underset{\sim}{P}$ is nonsingular

$$x^P = \underset{\sim}{P}^{-1} \underset{\sim}{x}$$

In terms of the transformation matrix $\underset{\sim}{P}$, the elements of the phase variable representation may be determined from the following equations.

$$\underset{\sim}{A}^P = \underset{\sim}{P}^{-1} \underset{\sim}{A} \underset{\sim}{P} \;, \quad \underset{\sim}{b}^P = \underset{\sim}{P}^{-1} \underset{\sim}{b} \;, \quad \underset{\sim}{c}^P = \underset{\sim}{P}^T \underset{\sim}{c}$$

In addition, the feedback coefficients in phase variables and the original variables are related by the following expressions.

$$\underset{\sim}{k}^P = \underset{\sim}{P}^T \underset{\sim}{k} \qquad \text{and} \qquad \underset{\sim}{k} = (\underset{\sim}{P}^T)^{-1} \underset{\sim}{k}^P$$

Therefore once the matrix $\underset{\sim}{P}$ is known, the complete transformation problem is solved.

If the coefficients of the characteristic polynomial of $\underset{\sim}{A}$, that is the a_i's are known[2], then $\underset{\sim}{P}$ may be determined by a simple algorithm[3] which indicates that if the vectors $\underset{\sim}{p}^i$ are defined by the recursion formula

$$\underset{\sim}{p}^n = \underset{\sim}{b}$$

and

$$\underset{\sim}{p}^{n-i} = \underset{\sim}{A} \underset{\sim}{p}^{n-i+1} + a_{n-i+1} \underset{\sim}{b} \qquad i = 1, 2, \ldots, n-1$$

then $\underset{\sim}{P}$ is given by

$$\underset{\sim}{P} = [\underset{\sim}{p}^1 \mid \underset{\sim}{p}^2 \mid \ldots \mid \underset{\sim}{p}^n]$$

This procedure is simple and well suited to use on a digital computer.

---

[1] R. E. Kalman, "Mathematical Description of Linear Dynamical Systems," J.S.I.A.M. Control, Ser. A, Vol. 1, no. 2, pp. 152-192, 1963.

[2] See Appendix A.

[3] W. G. Tuel, Jr., "On the Transformation to (Phase-Variable) Canonical Form," IEEE Trans. on Auto. Control, Vol. AC-11, no. 3 (July 1966), pp. 607. M. R. Chidambaram, "Comment on 'On the Transformation to (Phase-Variable) Canonical Form'," IEEE Trans. on Auto. Control, Vol. AC-11, no. 3 (July 1966), pp. 6-7-608. D. S. Rane, "A Simplified Transformation to (Phase-Variable) Canonical Form," IEEE Trans. on Auto. Control, Vol. AC-11, no. 3 (July 1966), p. 608. C. D. Johnson and W. M. Wonham, "Another Note on the Transformation to Canonical (Phase-Variable) Form," IEEE Trans. on Auto. Control, Vol. AC-11, no. 3 (July 1966), pp. 609-610.

Since the algorithm involves transforming the plant to phase variables, it is necessary that the plant be controllable. In addition, the plant must be controllable before any transfer function techniques can be sensibly utilized. Hence it is important to check the controllability of the plant to ensure that the results obtained are meaningful. Controllability should be checked even if the physical plant is known to be controllable since the mathematical modeling, especially if linearization has been involved, may have destroyed the property.

Controllability may be determined by the test, originally proposed by Kalman, which states that the plant

$$\dot{\underline{x}} = \underline{A}x + \underline{b}u$$

is controllable if and only if the n by n controllability matrix

$$\underline{M}_c = [\underline{b} \mid \underline{A}\underline{b} \mid \underline{A}^2\underline{b} \mid \ldots \mid \underline{A}^{n-1}\underline{b}]$$

is nonsingular, i.e. det $(\underline{M}_c) \neq 0$.

Even when the det $(\underline{M}_c) \neq 0$[1], problems may still arise if the matrix $\underline{M}_c$ is ill-conditioned and therefore difficult to invert. In order to check this possibility, the matrix $\underline{M}_c$ may be inverted, as well as possible, by the numerical inversion procedure to be used. If this inverse matrix $\underline{M}_c^{-1}$ is multiplied with $\underline{M}_c$, the result should be the identity matrix $\underline{I}$. The degree to which this product deviates from $\underline{I}$ is a measure of the uncontrollability of the plant. If the maximum value of absolute deviation is not negligible[2], the plant is identified as being <u>numerically</u> uncontrollable. This terminology is chosen to indicate that while the plant is theoretically controllable, it is uncontrollable in a numerical sense since $\underline{M}_c$ cannot be accurately inverted by the inversion algorithm. Although the numerical controllability of the plant is obviously dependent on the numerical inversion scheme used, a wide variety of methods for inversion have been tried with only minor changes in the results.

The computation algorithm permits a simple check to be made on the validity of the results obtained. This computational double-check is based on the fact that $D_k(s)$ may be determined by either of the following two expressions.

$$D_k(s) = D_p(s) + KN_h(s)$$

or

$$D_k(s) = \det(s\underline{I} - \underline{A} + K\underline{b}\underline{k}^T)$$

A normalized error is obtained by dividing the difference between the two values of a coefficient of $D_k(s)$ by the value of the coefficient. The maximum, over all of the coefficients of $D_k(s)$, of the absolute values of this normalized error is an indication of the accuracy of the closed-loop calculation. Because of the dissimilarities of the mathematical techniques in the two equations for $D_k(s)$ this procedure provides a simple and yet meaningful check on the accuracy of the results, which may be used to assist in determining the validity of the results in the numerically uncontrollable situation discussed above and to uncover any other numerical problems that might arise, although none have occured so far in the numerous examples that have been solved.

2.6-2 *INPUT FORMAT* The general input and output formats for the program are discussed in this and the following section respectively. While reading these sections, it is suggested that the reader refer often to the specific example problem presented in Sec. 2.6-4 especially Tables 2.6-7 and 2.6-8.

---

[1]Note that the exact value of det$(\underline{M}_c)$ is unimportant since it may be set at any value by scaling the original plant equations.

[2]A value larger than 10^{-3} to 10^{-5} has been found to indicate difficulty.

The operation of the program is divided into three phases: 1) basic, 2) open-loop and 3) closed-loop. During the basic phase, the A and b matrices are read and controllability is checked. The open-loop phase consists of the open-loop computations discussed in the previous sections and include the determination of $D_p(s)$, P and $N_p(s)$. The closed-loop phase varies depending on which types of closed-loop calculations are desired. In addition, the double check on accuracy is made during the closed-loop phase of operation. It is possible to execute a number of open- and closed-loop calculations as part of one basic problem as long as the plant is not changed.

The input and output formats fall into three similar divisions. Because of this fact and in order to simplify the treatment, the discussion of the input and output formats is divided along these same lines.

Basic Input: The input for the basic phase of the programs consists of a set of cards which specify the elements of A and b. In addition, a card is included which gives problem identification information for easy referencing of the problem and the order of the plant, n. These cards are included for all problems independent of what open- or closed-loop computations are to be made. A detailed description of the preparation of the basic input cards is shown in Table 2.6-1.

Table 2.6-1

Card Description For Basic Input

Card Number	Column Number	Description	Format
1	1-20	Problem identification	4A5,I2
	21-22	n = order of the plant	
2	1-10	a_{11}	8E10.0
	11-20	a_{12}	
	etc.	...	
3	1-10	a_{21}	8E10.0
	11-20	a_{22}	
	etc.	...	
n+2	1-10	b_1	8E10.0
	11-20	b_2	
	etc.	...	

Open-Loop Input: The input for the open-loop phase consists solely of c matrices which specify for which transfer functions the numerator polynomials are desired. The reader should recall that fictitious c matrices may be used to obtain internal transfer functions. The elements of each c matrix are placed on one or more cards, depending on the value of n, in the same manner as b, i.e. c_1 in column 1-10, c_2 in column 11-20, and so forth. (See Table 2.6-2).

43

Any number of $\underset{\sim}{c}$ matrices can be included in any order as long as two simple rules are followed.

Table 2.6-2

Card Description for the Open-Loop Input

(NOL = number of open-loop calculations desired)

Card Number	Column Number	Description	Format
1	1-10	c_1^1 = first element of the first fictitious c matrix	8E10.0
	11-20	c_2^1 = second element of the first fictitious c matrix	
	etc.	...	

. . . . Similarly for each remaining fictitious c matrices

NOL	1-10	c_1 = first element of the real c matrix	8E10.0
	11-20	c_2	
	etc.	...	
NOL+1	1-80	BLANK CARD	

First, the last $\underset{\sim}{c}$ matrix must be completely zero, i.e. $\underset{\sim}{c} = \underset{\sim}{0}$; this $\underset{\sim}{c}$ matrix of zeros is used only to signal the end of the open-loop calculations and no computations are made with it. Setting $\underset{\sim}{c}$ equal to zero may be easily accomplished by placing one or more blank cards in the deck depending on the value of n^1. The second rule is that the $\underset{\sim}{c}$ matrix for the real, physical output must always appear just before the blank $\underset{\sim}{c}$ matrix. This last rule is necessary in order that K be computed properly for zero steady state position error.

Closed-Loop Input: The program is capable of performing three types of closed-loop calculations which are indicated by setting the symbol KEY as A, P, or F respectively. One of these three computations (KEY = A) is for closed-loop analysis while the other two (KEY = P and F) are for design. Any number of these three types of computations in any combination may be accomplished at one time.

In the analysis mode, KEY = A, the program must be given K and $\underset{\sim}{k}$ as input. From this input the program determines the coefficients of the closed-loop characteristic polynomial, $D_k(s)$, and the numerator of $H_{eq}(s)$, $N_h(s)$.[2] In addition, both $D_k(s)$ and $N_h(s)$ are factored and their roots are displayed. This mode of operation is perhaps most useful for sensitivity studies. By varying the values of K and $\underset{\sim}{k}$, it is possible to see the effect of such changes

---

[1]For $n \leq 8$, one card is used; for $9 \leq n \leq 16$, two cards are used and so forth.

[2]The reader will recall that the denominator polynomial of $H_{eq}(s)$ is equal to the numerator of $G_p(s)$, i.e. $N_p(s)$.

on $y(s)/r(s)$. One may use this procedure for plotting the root locus of $1 + KG(s)H_{eq}(s)$ versus K. In addition, since $G_p(s)$ and $H_{eq}(s)$ are known in factored form, the root locus may be easily sketched by hand.

The input for the first of the two design modes, KEY = P, is the desired closed-loop characteristic polynomial, $D_k(s)$. From this information, the program computes K and $\underset{\sim}{k}$ and determines the numerator polynomial of $H_{eq}(s)$. In the second design mode, KEY = F, the input is the desired closed-loop pole locations and output is again K, $\underset{\sim}{k}$ and $H_{eq}(s)$. As in the analysis mode, the polynomial $D_k(s)$ and $N_h(s)$ are always given in factored as well as unfactored form. A summary of the three modes of closed-loop calculations is given in Table 2.6-3.

Table 2.6-3

Summary of Closed-Loop Computations

KEY	TYPE	INPUT	OUTPUT
A	Analysis	K and $\underset{\sim}{k}$	$N_h(s)$ and $D_k(s)$
P	Design	$D_k(s)$, unfactored	$N_h(s)$, K and $\underset{\sim}{k}$
F	Design	$D_k(s)$, factored	$N_h(s)$, K and $\underset{\sim}{k}$

The reader is reminded that K is always selected so that zero steady-state error for a step input results. If it is desired that some other condition be used to select K, then one may always rescale K and $\underset{\sim}{k}$ to meet such a restriction by hand. Suppose, for example, that is desired that the d.c. gain be ten rather than one, then one would simply multiply K by ten and divide $\underset{\sim}{k}$ by ten. This procedure leaves the denominator of $y(s)/r(s)$ unchanged but multiplies the numerator by ten as desired.

Each closed-loop calculation is represented by a set of two or more cards whose format depends on which of the three types of closed-loop calculations is desired. (See Tables 2.6-4, -5, and -6). These sets, one for each calculation, in any order and of any number, are combined to form the input for the closed-loop phase of the program. For example, if one wished to make three analysis mode (KEY = A) and two second design mode (KEY = F) calculations, then he would make up five sets of cards, one for each calculation, using the format appropriate for each calculation. These cards would then be combined by placing the three sets of the analyzing calculations first, the two design sets first, or the analysis and design sets could be intermixed in any order as long as the individual sets remained together as units.

Note that the first card in the format for each type of closed-loop calculation has only one letter which is punched in the first column to give the value of KEY. The remaining cards in each set depend on the input that is required for the desired type of calculation.

A single blank card is placed in the input deck after the desired closed-loop calculation sets to signal the end of the closed-loop input. When this blank card is read, the program finds that the value of KEY is zero and returns to the basic input section to find if another problem remains to be solved. The program is stopped by the end-of-file indicator. Even if no closed-loop calculations are desired, this blank card must still be included. In this case, there are two or more blank cards at the end of the input deck since one or more blank cards appear at the end of the open loop input.

Table 2.6-4

Card Description for the Closed-Loop Input-Analysis Model
(KEY = A)

Card Number	Column Number	Description	Format
1	1	KEY = A	A1
2	1-10	K = Forward path gain	E10.0
3	1-10	k_1	8E10.0
	11-20	k_2	
	etc.	...	

Table 2.6-5

Card Description for the Closed-Loop Input-First Design Mode
(KEY = P)

Card Number	Column Number	Description	Format
1	1	KEY = P	A1
2	1-10	a_{k1}	8E10.0
	11-20	a_{k2}	
	etc.[1]	...	

[1]Note that the coefficient a_{kn+1} of s^n is always assumed to be one and is <u>not</u> included on the input cards.

Table 2.6-6

Card Description for the Closed-Loop Input-Second Design Mode
(KEY = F)

Card Number	Column Number	Description	Format
1	1	KEY = F	A1
2	1-10	real part of the first pole of the desired $y(s)/r(s)$	2E10.0
	11-20	imaginary part of the first pole of the desired $y(s)/r(s)$	
3	1-10	real part of the second pole of the desired $y(s)/r(s)$	2E10.0
	11-20	imaginary part of the second pole of the desired $y(s)/r(s)$	
etc.	...	...	

In this section, the method of preparing the input deck for state variable feedback program has been discussed in detail. Sec. 2.6-4 contains an example illustrating the preparation of the input deck for a specific problem.

2.6-3 *OUTPUT INTERPRETATION*   The output format is divided into the same three phases as the input, viz. basic, open-loop and closed-loop. The reader is urged to refer to the specific example discussed in Sec. 2.6-4 especially Table 2.6-8 while reading this section.

Basic Output: The basic output is composed, in the main, of simply a printout of the information contained in the basic input, i.e. the alphanumeric problem identification, and the $\underline{A}$ and $\underline{b}$ matrices. This material is provided solely as reference information for the user since no new results are included. For most problems, these items are the entire basic output. However, for problems which are either uncontrollable or numerically uncontrollable,[1] an additional item is added to the basic output in order to indicate this complication. The reader should remember that even if the plant is found to be uncontrollable that all open- and closed-loop calculations are still performed. In the case of numerical uncontrollability, the maximum deviation of the matrix $\underline{M}_c \underline{M}_c^{-1}$ from the identity matrix is also listed for the information of the user.

Open-Loop Output: The open-loop output is composed of the denominator of the plant transfer function, $D_p(s)$, and the various numerator polynomials associated with the plant transfer function and any internal transfer functions desired. The coefficients of $D_p(s)$, as well as those of all other polynomials, are listed with the <u>coefficient of the constant term first</u>. In addition to the coefficients of $D_p(s)$, the roots of $D_p(s)$ are also listed. The various numerator coefficients are identified by printing the $\underline{c}$ matrix associated with each. Each numerator polynomial is listed in its factored as well as its unfactored form.

Closed-Loop Output: The output of the closed-loop phase of the program has exactly the same form for each of the three possible types of closed-loop calculations. The first item in the closed-loop output, however, is the value of KEY which identifies the type of computation performed. The next four items in the output are the numerator polynomial of $H_{eq}(s)$, $N_h(s)$, the feedback coefficients $\underline{k}$, the gain K, and the denominator of $y(s)/r(s)$, $D_k(s)$. Both $N_h(s)$ and $D_k(s)$ are given in both factored and unfactored form.

The last item in the closed-loop output is the value of the maximum normalized error.[2] The reader will remember that this number is an indication of the numerical accuracy of the program and is obtained by comparing the matrix and transfer function methods for finding $D_k(s)$. If more than one closed-loop computation has been requested, then the output discussed is listed for each computation.

2.6-4 *EXAMPLE PROBLEM*   This subsection discusses the application of the state variable feedback program to a simple, but not trivial, third-order example. This example is presented in order to clarify the details of preparing the input deck for interpreting the printout of the program. It is suggested that the reader refer often to the previous sections while reading this section in order to achieve maximum understanding.

The state variable representation of the plant for the example is given by

$$\dot{\underline{x}}(t) = \begin{bmatrix} -1.0 & 1.0 & 0 \\ 0 & 0 & 1.0 \\ 0 & -3.0 & 0 \end{bmatrix} \underline{x}(t) + \begin{bmatrix} 0 \\ 0 \\ 1.0 \end{bmatrix} u(t)$$

$$y(t) = [1.0 \quad 1.0 \quad 0]\underline{x}(t)$$

---

[1] See Sec. 2.6-1 for a definition of numerical uncontrollability.

[2] See Sec. 2.6-1 for a complete discussion.

The block diagram of the plant is shown in Fig. 2.6-1. From this information it is possible to prepare the basic input portion of the input deck. (See Table 2.6-7) This information must be provided no matter what open- or closed-loop calculations are to be performed.

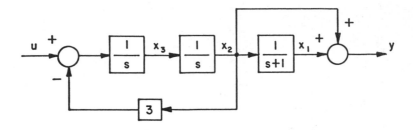

Fig. 2.6-1 Block Diagram of the Plant

In terms of open-loop calculations, it is desired to obtain the internal transfer function $x_3(s)/u(s)$ in addition to the plant transfer function $y(s)/u(s)$. Therefore it is necessary to form a fictitious $\underset{\sim}{c}$ matrix which may be used to find this internal transfer function which is given by

$$\underset{\sim}{c}^1 = \text{col } (0, 0, 1.0)$$

and the open-loop input takes the form shown in Table 2.6-7. Note that the real $\underset{\sim}{c}$ matrix, $\underset{\sim}{c} = \text{col } (1.0, 1.0, 0)$, is placed just before the blank card indicating the end of the open-loop input.

In a practical problem, one would normally run the program one time with no closed-loop calculations, i.e. only a blank card for the closed-loop input, in order to have the open-loop information available for the specification of $y(s)/r(s)$, which has been chosen to be

$$\frac{y(s)}{r(s)} = \frac{s(s + 2)}{[(s + 1)^2 + 1^2](s + 2)} = \frac{2(s + 2)}{(s^3 + 4s^2 + 6s + 4)}$$

Since $D_k(s)$ is known in both factored and unfactored form, either of the closed-loop design modes could be used to find K and $\underset{\sim}{k}$. In order to illustrate the preparation of the input deck, both of the design modes are used. In addition, the closed-loop analysis mode is utilized by letting

$$K = 2.0 \quad \text{and} \quad \underset{\sim}{k} = \text{col}(0.5, 0, 1.5)$$

These values for K and $\underset{\sim}{k}$ are selected, with the advantage of foresight, so that they yield the desired $y(s)/r(s)$. The complete input deck is shown in Table 2.6-7.

This problem is obviously somewhat artificial since both of the design modes are used when only one is necessary and the analysis mode is being run at the same time. The use of all three of the closed-loop modes is strictly for illustrative purposes and should not be construed as a typical utilization of the program.

The complete output for this problem is shown in Table 2.6-8. Since there is no indication of uncontrollability in the basic output, it can be safely assumed that the plant is controllable and therefore the use of the program is justified. The open-loop portion of the output provides all of the desired open-loop transfer function. The plant transfer function is seen to be

Table 2.6-7

Input Deck for STVARFDBK

Card No.	Column No. 1-5	5-10	10-15	15-20	20-25	25-30
1	EXAMPLE ONE				3	
2	-1.0		1.0		0.0	
3	0.0		0.0		1.0	
4	0.0		-3.0		0.0	
5	0.0		0.0		1.0	
6	0.0		0.0		1.0	
7	1.0		1.0		0.0	
8	(BLANK CARD)					
9	A					
10	2.0					
11	0.5		0.0		1.5	
12	P					
13	4.0		6.0		4.0	
14	F					
15	1.0		1.0			
16	2.0					
17	(BLANK CARD)					

$$G(s) = \frac{s + 2}{s^3 + s^2 + 3s + 3} = \frac{s + 2}{[s^2 + (\sqrt{3})^2](s + 1)}$$

while the internal transfer function, $x_3(s)/u(s)$, for example is

$$\frac{x_3(s)}{u(s)} = \frac{s^2 + s}{s^3 + s^2 + 3s + 3} = \frac{s}{[s^2 + (\sqrt{3})^2]}$$

The closed-loop output of the two design modes (KEY = P and F) indicates that the K and k necessary to produce the desired $y(s)/r(s)$ are

$$K = 2.0 \quad \text{and} \quad \underset{\sim}{k} = \text{col}(0.5, 0, 1.5)$$

The output of the analysis mode, on the other hand, confirms that these values for K and $\underset{\sim}{k}$ do, in fact, provide the desired results. The maximum normalized error for each of the design modes is extremely small indicating that the accuracy of the results is acceptable.

49

Table 2.6-8

Program Output for STVARFDBK Example

```
STATE VARIABLE FEEDBACK
PROBLEM IDENTIFICATION -      EXAMPLE ONE

*******************************************

THE A MATRIX

-1.0000000E 00          1.0000000E 00          0.0
 0.0                    0.0                    1.0000000E 00
 0.0                   -3.0000000E 00          0.0

THE B MATRIX

 0.0                    0.0                    1.0000000E 00

*******************************************

OPEN-LOOP CALCULATIONS

DEMONINATOR COEFFICIENTS - IN ASCENDING POWERS OF S

 2.9999990E 00          3.0000000E 00          1.0000000E 00          1.00

THE ROOTS ARE                   REAL PART               IMAGINARY PART
                                -0.0                    -1.7320499E 00
                                -0.0                     1.7320499E 00
                                -1.0000000E 00           0.0


THE C MATRIX      *****

 0.0                    0.0                    1.0000000E 00

NUMERATOR COEFFICIENTS - IN ASCENDING POWERS OF S

 0.0                    1.0000000E 00          1.0000000E 00

THE ROOTS ARE                   REAL PART               IMAGINARY PART
                                -1.0000000E 00           0.0
                                 0.0                     0.0


THE C MATRIX      *****

 1.0000000E 00          1.0000000E 00          0.0

NUMERATOR COEFFICIENTS - IN ASCENDING POWERS OF S

 2.0000000E 00          1.0000000E 00

THE ROOTS ARE                   REAL PART               IMAGINARY PART
                                -2.0000000E 00           0.0

:*******************************************
```

Table 2.6-8 (cont.)

CLOSED—LOOP CALCULATIONS

KEY = A   *****

THE NUMERATOR OF H—EQUIVALENT — IN ASCENDING POWERS OF S

 5.0000000E-01   1.5000000E 00   1.5000000E 00

THE ROOTS ARE   REAL PART   IMAGINARY PART
                       -5.0000000E-01   -2.8867507E-01
                       -5.0000000E-01   2.8867507E-01

THE FEEDBACK COEFFICIENTS

 5.0000000E-01   0.0   1.5000000E 00

THE GAIN =   2.0000000E 00

THE CLOSED—LOOP CHARACTERISTIC POLYNOMIAL — IN ASCENDING POWERS (

 3.9999990E 00   6.0000000E 00   4.0000000E 00   1.0(

THE ROOTS ARE   REAL PART   IMAGINARY PART
                       -1.0000000E 00   -1.0000000E 00
                       -1.0000000E 00   1.0000000E 00
                       -1.9999990E 00   0.0

MAXIMUM NORMALIZED ERROR =   0.0

KEY = P   *****

THE NUMERATOR OF H—EQUIVALENT — IN ASCENDING POWERS OF S

 5.0000048E-01   1.5000000E 00   1.5000000E 00

THE ROOTS ARE   REAL PART   IMAGINARY PART
                       -5.0000000E-01   -2.8867561E-01
                       -5.0000000E-01   2.8867561E-01

THE FEEDBACK COEFFICIENTS

 5.0000048E-01   0.0   1.5000000E 00

THE GAIN =   2.0000000E 00

THE CLOSED—LOOP CHARACTERISTIC POLYNOMIAL — IN ASCENDING POWERS (

 4.0000000E 00   6.0000000E 00   4.0000000E 00   1.0(

THE ROOTS ARE   REAL PART   IMAGINARY PART
                       -1.0000000E 00   -1.0000000E 00
                       -1.0000000E 00   1.0000000E 00
                       -1.9999990E 00   0.0

MAXIMUM NORMALIZED ERROR =   0.0

Table 2.6-8 (cont.)

KEY = F   *****

THE NUMERATOR OF H-EQUIVALENT - IN ASCENDING POWERS OF S

5.0000024E-01   1.5000000E 00   1.5000000E 00

THE ROOTS ARE   REAL PART   IMAGINARY PART
                            -5.0000000E-01   -2.8867537E-01
                            -5.0000000E-01   2.8867537E-01

THE FEEDBACK COEFFICIENTS

5.0000024E-01   0.0   1.5000000E 00

THE GAIN =   1.9999990E 00

THE CLOSED-LOOP CHARACTERISTIC POLYNOMIAL - IN ASCENDING POWERS O

3.9999990E 00   6.0000000E 00   4.0000000E 00   1.00

THE ROOTS ARE   REAL PART   IMAGINARY PART
                            -1.0000000E 00   -1.0000000E 00
                            -1.0000000E 00   1.0000000E 00
                            -1.9999990E 00   0.0

MAXIMUM NORMALIZED ERROR =   4.77E-07

    2.6-5 *SUBPROGRAMS USED*   The following subprograms are used by this
program:

        (1)   CHREQA
        (2)   PROOT
        (3)   SIMEQ

See Appendix A for a description of each of these subprograms.

    2.6-6 *PROGRAM LISTING*

```
C   STATE VARIABLE FEEDBACK PROGRAM (STVARFDBK)
C   SUBPROGRAMS USED  SIMEQ, PROOT, CHREQA.
        DOUBLE PRECISION NAME
        DIMENSION  A(10,10),B(10),H(10),C(10),NAME(4),AA(10,10)
        DIMENSION D(10),P(10,10),CC(10),HH(10),PIN(10,10),U(10)
        DIMENSION E(10),V(10),LETTER(4)
        DATA LETTER(1),LETTER(2),LETTER(3),LETTER(4)/
     *     1H ,1HA,1HP,1HF/
 2000 FORMAT (1H0,45(1H*))
 2001 FORMAT (4A5,I2)
 2002 FORMAT(8E10.0)
 2003 FORMAT (     6X,25HPROBLEM IDENTIFICATION -    ,5X,4A5)
 2004 FORMAT(1H0,5X,12HTHE A MATRIX/)
 2005 FORMAT(1P6E20.7)
 2006 FORMAT(1H0,5X,12HTHE B MATRIX/)
 2007 FORMAT(1H0,5X,41HTHE CLOSED-LOOP CHARACTERISTIC POLYNOMIAL
     *     ,27H - IN ASCENDING POWERS OF S    /)
 2008 FORMAT(1H0,5X,25HTHE FEEDBACK COEFFICIENTS/)
 2009 FORMAT(1H0,5X,10HTHE GAIN =1PE16.7)
 2010 FORMAT(1H0,5X,12HTHE C MATRIX,5X,5H*****/)
 2011 FORMAT(1H0,5X,24HDEMONINATOR COEFFICIENTS
     *     ,27H - IN ASCENDING POWERS OF S    /)
```

52

```
  2012 FORMAT(1H0,5X,22HNUMERATOR COEFFICIENTS
     *    ,27H - IN ASCENDING POWERS OF S     /)
  2013 FORMAT(1H0,5X,29HTHE NUMERATOR OF H-EQUIVALENT
     *    ,27H - IN ASCENDING POWERS OF S     /)
  2014 FORMAT(1H0,5X,22HOPEN-LOOP CALCULATIONS)
  2015 FORMAT(1H0,5X,26HMAXIMUM NORMALIZED ERROR = 1PE10.2/)
  2016 FORMAT (A1)
  2017 FORMAT(1H0,5X,24HCLOSED-LOOP CALCULATIONS)
  2018 FORMAT (1H0,5X,6HKEY = A1,3X,5H*****)
  2019 FORMAT(1H0,5X, 23HPLANT IS UNCONTROLLABLE5X,10H**********)
  2020 FORMAT(1H0,4X,  35HPLANT IS NUMERICALLY UNCONTROLLABLE10X,
     1 16HMAX. DEVIATION = 1PE10.2,5X,10H**********)
  2021 FORMAT (1H0,5X,14HTHE ROOTS ARE ,13X,9HREAL PART,10X,
     *    14HIMAGINARY PART)
  2022 FORMAT(25X,1P2E20.7)
  2023 FORMAT(1H0)
  2024 FORMAT(1H1,5X,24HSTATE VARIABLE FEEDBACK    )
C     READ INPUT DATA
       NRM=1
     1 READ (5,2001,END=10) (NAME(I),I=1,4), N
       PRINT 2024
  1111 PRINT 2003,(NAME(I),I=1,4)
       PRINT 2000
       PRINT 2004
       DO 2 I=1,N
       READ 2002,(A(I,J),J=1,N)
     2 PRINT 2005,(A(I,J),J=1,N)
       PRINT 2006
       READ 2002,(B(I),I=1,N)
       PRINT 2005,(B(I),I=1,N)
       NKEY=0
C     CHECK CONTROLLABILITY
       DO 7 I=1,N
     7 AA(I,1)=B(I)
       DO 8 I=2,N
       L=I-1
       DO 8 J=1,N
       AA(J,I)=0.
       DO 8 K=1,N
     8 AA(J,I)=AA(J,I)+A(J,K)*AA(K,L)
       CALL SIMEQ (AA,C,N,PIN,C,ICNTR)
       IF(ICNTR) 3,4,3
     4 PRINT 2019
       GO TO 9
     3 DO 43 I=1,N
       DO 43 J=1,N
       P(I,J)=0.
       DO 43 K=1,N
    43 P(I,J)=P(I,J)+AA(I,K)*PIN(K,J)
       ERROR=0.
       DO 44 I=1,N
       DO 44 J=1,N
       IF(I-J) 45,46,45
    46 ERR = ABS (P(I,J)-1.0)
       GO TO 44
    45 ERR = ABS (P(I,J))
    44 ERROR=AMAX1(ERROR,ERR)
       IF(ERROR-1.E-5) 9,47,47
    47 PRINT 2020,ERROR
C     OPEN-LOOP CALCULATIONS
     9 PRINT 2000
       PRINT 2014
       NN=N+1
```

53

```
      PRINT 2011
      CALL CHREGA(A,N,D)
      PRINT 2005,(D(I),I=1,NN)
      CALL PROOT(N,D,U,V,+1)
      PRINT 2021
      PRINT 2022,(U(I),V(I),I=1,N)
      DO11 I=1,N
   11 P(I,N)=B(I)
      DO12JJ=2,N
      DO12 I=1,N
      J=N-JJ+1
      K=J+1
      P(I,J)=D(K)*B(I)
      DO12 L=1,N
   12 P(I,J)=P(I,J)+A(I,L)*P(L,K)
   72 READ 2002,(C(I),I=1,N)
      DO 70 I=1,N
      IF(C(I)) 71,70,71
   70 CONTINUE
      GO TO 104
   71 PRINT 2023
      PRINT 2010
      PRINT 2005,(C(I),I=1,N)
   49 DO13 I=1,N
      CC(I)=0.
      DO13 J=1,N
   13 CC(I)=P(J,I)*C(J)+CC(I)
      DO 100 I=1,N
      M=NN-I
      IF(CC(M))101,100,101
  100 CONTINUE
  101 PRINT 2012
      PRINT 2005,(CC(I),I=1,M)
      M=M-1
      IF(M) 105,105,103
  103 CALL PROOT(M,CC,U,V,+1)
      PRINT 2021
      PRINT 2022,(U(I),V(I),I=1,M)
  105 GO TO 72
C        CLOSED-LOOP CALCULATIONS
  104 READ 2016, KEYX
      DO 80 I=1,4
      IF (LETTER(I).EQ.KEYX) GO TO 81
   80 CONTINUE
   81 KEY = I-1
      IF(KEY)40,10,40
   40 IF(NKEY) 120,120,121
  120 NKEY=1
      PRINT 2000
      PRINT 2017
  121 PRINT 2018, KEYX
      GO TO (21,20,22),KEY
   21 READ 2002,GAIN
      READ 2002,(HH(I),I=1,N)
      DO 41 I=1,N
      H(I)=0.
      DO 41 J=1,N
   41 H(I)=H(I)+P(J,I)*HH(J)
      DO 42 I=1,N
      DO 42 J=1,N
   42 AA(I,J)=A(I,J)-GAIN*B(I)*HH(J)
      CALL CHREGA(AA,N,E)
      GO TO 24
```

```
   22 DØ 23 I=1,N
      DØ 23 J=1,N
   23 AA(I,J)=0.
      I=0
   34 I=I+1
      READ 2002,RR,RI
      IF(RI) 30,31,30
   31 AA(I,I)=-RR
      GØ TØ 35
   30 AA(I,I)=-RR
      J=I+1
      AA(I,J)=-RI
      AA(J,I)=RI
      I=J
      AA(I,I)=-RR
   35 IF(I-N) 34,50,50
   50 CALL CHREGA(AA,N,E)
      GØ TØ 25
   20 READ 2002,(E(I),I=1,N)
   25 DØ19 I=1,N
   19 H(I)=E(I)-D(I)
      CALL SIMEG(P,C,N,PIN,C,ER)
   18 DØ16 I=1,N
      HH(I)=0.
      DØ16 J=1,N
   16 HH(I)=HH(I)+PIN(J,I)*H(J)
      DØ 5 I=1,N
      DØ 5 J=1,N
    5 AA(I,J)=A(I,J)-B(I)*HH(J)
      CALL CHREGA(AA,N,E)
      GAIN =E(1)/CC(1)
      DØ26 I=1,N
      HH(I)=HH(I)/GAIN
   26 H(I)=H(I)/GAIN
   24 PRINT 2013
      PRINT 2005,(H(I),I=1,N)
      N1=N-1
      CALL PRØØT(N1,H,U,V,+1)
      PRINT 2021
      PRINT 2022,(U(I),V(I),I=1,N1)
   17 PRINT 2008
      PRINT 2005,(HH(I),I=1,N)
      PRINT 2009,GAIN
      PRINT 2007
      PRINT 2005,(E(I),I=1,NN)
      CALL PRØØT(N,E,U,V,+1)
      PRINT 2021
      PRINT 2022,(U(I),V(I),I=1,N)
C     CHECK ACCURACY
      ERRØR =0.
      DØ 6 I=1,N
      ERR=(E(I)-GAIN*H(I)-D(I))/E(I)
      ERR= ABS (ERR)
    6 ERRØR=AMAX1(ERRØR,ERR)
  106 PRINT 2015,ERRØR
      GØ TØ 104
   10 STOP
      END
```

2.7   OBSERVABILITY INDEX CALCULATION (OBSERV)

The OBSERV program is used to determine the observability index of the system

$$\dot{x}(t) = Ax(t) \qquad\qquad (2.7\text{-}1)$$

$$y(t) = Cx(t) \qquad\qquad (2.7\text{-}2)$$

where $x(t)$ is the n-dimensional state vector and $y(t)$ is the m-dimensional output measurement vector. The observability index is used in the programs SERCOM and LUEN to determine the order of the compensator needed.

2.7-1   *THEORY*   The observability index of the system of Eqs. (2.7-1) and (2.7-2) is the minimum integer ν such that the matrix Q_ν defined by

$$Q_\nu = [C^T \mid A^T C^T \mid \cdots \mid [A^T]^{(\nu-1)} C^T]$$

has rank n. In general $n/m \leq \nu \leq n - m + 1$. The matrix Q_i, $i = 1,2,\ldots$ is calculated and its rank is determined by the use of HERMIT. If the rank of Q_j is n while the rank of Q_{j-1} is less than n, then the observability index ν is equal to j. If the rank of Q_n is less than n, then the system is labeled as unobservable.

2.7-2   *INPUT FORMAT*   The first input card contains the problem identification and system order information material. In addition, the first card indicates the dimension of y. The next N cards give the elements of A in the usual format; the following M cards give the rows of C in the usual manner. The input format is summarized in Table 2.7-1.

2.7-3   *OUTPUT FORMAT*   For reference, the problem identification as well as A and C are printed, in the usual form, at the beginning of the output. The remaining output will vary depending on the observability characteristics of (A,C). If the system is unobservable the message

(A,C) IS UNOBSERVABLE

is printed. If (A,C) is observable with index I, then the message is

OBSERVABILITY INDEX = I

2.7-4   *EXAMPLE*   Let us determine the observability index for the system

$$\dot{x}(t) = \begin{bmatrix} 1 & 0 & 0 & 0 \\ 0 & 2 & 0 & 0 \\ 0 & 0 & 3 & 0 \\ 0 & 0 & 0 & 4 \end{bmatrix} x(t)$$

$$y(t) = \begin{bmatrix} 1 & 2 & 0 & 0 \\ 0 & 0 & 3 & 0 \end{bmatrix} x(t)$$

In this case N = 4, M = 2. The complete input deck for this example is shown in Table 2.7-2. The related output is given in Table 2.7-3.

TABLE 2.7-1

Input Format for OBSERV Program

Card Number	Column Number	Description	Format
1	1-20	Problem identification	5A4,2I2
	21-22	N = system order, $N \leq 10$	
	23-24	M = number of outputs, $M \leq 10$	
2	1-10	a_{11} = Plant matrix	8F10.3
	11-20	a_{12}	
	etc.	$\cdots$	
3	1-10	a_{21}	8F10.3
	11-20	a_{22}	
	etc.	$\cdots$	
N+2	1-10	c_{11}	8F10.3
	11-20	c_{12}	
	etc.	$\cdots$	
N+M+1	1-10	c_{m1}	8F10.3
	11-20	c_{m2}	

Table 2.7-2

Input Data Deck for OBSERV Example

Card No.	Column No.
1	NO. 4:UNOBSERVABLE 4 2
2	1.0 0 0 0
3	0 2.0 0 0
4	0 0 3.0 0
5	0 0 0 4.
6	1.0 2.0 0 0
7	0 0 3.0 0

Table 2.7-3

Program Output for OBSERV Example

OBSERVABILITY INDEX CALCULATION
PROBLEM IDENTIFICATION'   NO. 4:UNOBSERVABLE

****

THE A MATRIX

```
    1,0000000+00        -0.0000000        -0.0000000        -0,0
   -0,0000000         2,0000000+00        -0.0000000        -0,0
   -0,0000000        -0.0000000         3,0000000+00        -0,0
   -0,0000000        -0.0000000        -0.0000000         4,0
```

THE C MATRIX

```
    1,0000000+00        2,0000000+00        -0.0000000        -0,0
   -0,0000000        -0.0000000         3,0000000+00         0,0
```

****

(A,C) IS UNOBSERVABLE

****

2.7-5 *SUBPROGRAMS USED*   The following subprograms are used by this
program

      (1)   HERMIT

      (2)   MULT

See Appendix A for a description of each of these subprograms.

2.7-6 *PROGRAM LISTING*

```
      SUBROUTINE OBSERV
C   OBSERVABILITY INDEX PROGRAM
C     SUBROUTINES USED-- HERMIT, MULT
      DIMENSION A(10,10),C(10,10),P(20,10),NAME(5),SKL(10,10)
 1000 FORMAT (5A4,2I2)
 1001 FORMAT (1H1,5X,31HOBSERVABILITY INDEX CALCULATION/6X,
     * 23HPROBLEM IDENTIFICATION',5X,5A4)
 1002 FORMAT (1H0,45(1H*))
 1003 FORMAT (1H0,5X,12HTHE A MATRIX/)
 1004 FORMAT (8F10,3)
 1005 FORMAT (1P6E20,7)
 1006 FORMAT (1H0,5X,12HTHE C MATRIX/)
 1008 FORMAT (1H0,5X,21H(A,C) IS UNOBSERVABLE)
  100 READ(5,1000)(NAME(I),I=1,5),N,M
      PRINT 1001,(NAME(I),I=1,5)
      PRINT 1002
      PRINT 1003
      DO 10 I=1,N
      READ 1004,(A(I,J),J=1,N)
```

58

```
   10 PRINT 1005,(A(I,J),J=1,N)
      PRINT 1006
      DO 20 I=1,M
      READ 1004,(C(I,J),J=1,N)
   20 PRINT 1005,(C(I,J),J=1,N)
      PRINT 1002
      MM = M
      NU = 1
      DO 30 I=1,M
      DO 30 J=1,N
      P(I,J) = C(I,J)
   30 SKL(I,J) = C(I,J)
  200 CALL HERMIT (P,MM,N,IRANK)
      IF (IRANK.LT.N) GO TO 210
      PRINT 1007,NU
 1007 FORMAT(1H0,5X,22HOBSERVABILITY INDEX = ,I2)
      RETURN
  210 IF(NU.LT.N) GO TO 220
      PRINT 1008
      RETURN
  220 NU = NU+1
      CALL MULT (SKL,A,SKL,M,N,N)
      DO 230 I=1,M
      II = IRANK+I
      DO 230 J=1,N
  230 P(II,J) = SKL(I,J)
      MM = IRANK+M
      GO TO 200
      END
```

2.8 LUENBERGER OBSERVER DESIGN (LUEN)

The program LUEN is used to design Luenberger observers to achieve a given closed-loop transfer function when some state variables are inaccessible.

2.8-1 *THEORY*   It is assumed that the reader is familiar with the basic theory of the Luenberger Observer in the design of linear state variable feedback systems with inaccessible states[1]. The plant representation will take the form

$$\dot{x}(t) = A x(t) + b u(t) \tag{2.8-1}$$

$$y^*(t) = c^T x(t) \tag{2.8-2}$$

$$y(t) = C x(t) \tag{2.8-3}$$

Here $y^*(t)$ is the output variable to be controlled while $y(t)$ represents the measurable outputs. Normally $y^*(t)$ would be a member of $y(t)$ but this is not necessary. We wish to achieve a desired $y^*(s)/r(s)$ by the use of only the measurable output $y(t)$.

The approach of the Luenberger observer is to place a compensator in the feedback path which can be used to generate information about the unmeasurable states. The general system then takes the form shown in Fig. 2.8-1.

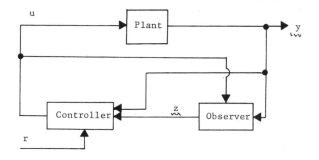

Fig. 2.8-1   Observer Form

The observer or feedback compensator will be described by

$$\dot{z}(t) = F z(t) + G_1 y(t) + G_2 u(t) \tag{2.8-4}$$

We wish to design the compensator such that we can write

$$k^T x = h^T z + g^T y \tag{2.8-5}$$

In order to do this we will force $z(t)$ to be a linear transformation of $x(t)$ so that

---

[1] See for example: D. G. Luenberger: "Observing the State of a Linear System," IEEE Trans. on Military Elect., vol. MIL-8, pp. 74-80, April 1964; D. G. Luenberger: "Observers for Multivariable Systems," IEEE Trans. on Auto. Control, vol. AC-11, pp. 190-197, April 1966; J. W. Weinstein and J. L. Melsa: "Design of Simplified Vertical Channel Landing Condition Autopilot Using State Variables," 1970 SWIEEECO Record, Dallas, Texas, April 22-24, 1970.

$$\underline{z}(t) = \underline{T}\underline{x}(t)$$

If $\underline{z}(0) = \underline{T}\underline{x}(0)$, then this condition will be satisfied if

$$\underline{\dot{z}}(t) = \underline{T}\underline{\dot{x}}(t)$$

or

$$\underline{F}\underline{z} + \underline{G}_1\underline{y} + \underline{G}_2 u = \underline{T}\underline{A}\underline{x} + \underline{T}\underline{b}u$$

But $\underline{z} = \underline{T}\underline{x}$ and $\underline{y} = \underline{C}\underline{x}$ so that we have

$$\underline{F}\underline{T}\underline{x} + \underline{G}_1\underline{C}\underline{x} + \underline{G}_2 u = \underline{T}\underline{A}\underline{x} + \underline{T}\underline{b}u \tag{2.8-6}$$

If this relation is to be satisfied for every $\underline{x}$ and u, we see that it is necessary that

$$\underline{F}\underline{T} + \underline{G}_1\underline{C} = \underline{T}\underline{A} \tag{2.8-7}$$

and

$$\underline{G}_2 = \underline{T}\underline{b} \tag{2.8-8}$$

In addition, we wish Eq. (2.8-5) to be satisfied so that

$$\underline{k}^T\underline{x} = \underline{h}^T\underline{T}\underline{x} + \underline{g}^T\underline{C}\underline{x}$$

and since this must also be valid for any $\underline{x}$

$$\underline{k}^T = \underline{h}^T\underline{T} + \underline{g}^T\underline{C} \tag{2.8-9}$$

Eqs. (2.8-7), (2.8-8) and (2.8-9) will always have at least one solution (and in general an infinite number) if:

(1) The eigenvalues of $\underline{F}$ do not equal any of the eigenvalues of $\underline{A}$.

(2) The dimension of $\underline{z}$ is equal to or greater than $\nu-1$, where the observability index is defined as the minimum integer such that $\underline{Q}_\nu$ is defined as

$$\underline{Q}_\nu = [\underline{C}^T \mid \underline{A}^T\underline{C}^T \mid \cdots \mid (\underline{A}^{\nu-1})^T\underline{C}^T]$$

has rank n. The observability index may be calculated by the use of the program OBSERV.

The complete design procedure now takes the following form.

(1) Select $y^*(s)/r(s)$ and solve for K and $\underline{k}$ as usual by the use of STVARFDBK.

(2) Determine ν by the use of OBSERV and select $\underset{\sim}{F}$ to satisfy the above properties.

(3) Solve Eqs. (2.8-7), (2.8-8) and (2.8-9) to obtain $\underset{\sim}{G}_1$, $\underset{\sim}{G}_2$, $\underset{\sim}{h}$ and $\underset{\sim}{g}$. This is the basic purpose of the program.

(4) The final system is then given by

$$\underset{\sim}{\dot{x}} = \underset{\sim}{A}x + \underset{\sim}{b}u \qquad (2.8\text{-}10)$$

$$\underset{\sim}{\dot{z}} = \underset{\sim}{F}z + \underset{\sim}{G}_1\underset{\sim}{y} + \underset{\sim}{G}_2 u \qquad (2.8\text{-}11)$$

$$u = K[r - \underset{\sim}{h}^T z - \underset{\sim}{g}^T y] \qquad (2.8\text{-}12)$$

$$y^* = \underset{\sim}{c}^T x \qquad (2.8\text{-}13)$$

$$\underset{\sim}{y} = Cx \qquad (2.8\text{-}14)$$

2.8-2 *INPUT FORMAT*   The first input card contains the problem identification, plant order N, dimension of the output vector M and the order of the observer L.   The next N + M + 1 cards consist of $\underset{\sim}{A}$, $\underset{\sim}{b}$, $\underset{\sim}{C}$ and $\underset{\sim}{k}$ in the usual format.   The $\underset{\sim}{F}$ matrix is assumed to be in phase variable form and is specified by giving its characteristic polynomial.[1] The polynomial may be supplied as either polynomial coefficients (OPTION = P) or as polynomial factors (OPTION = F) by placing either a P or F in column 1 of card N + M + 4.   Note that if the coefficient mode is selected, the coefficients of the highest order term is assumed to be unity and is not entered.   The complete input format is summarized in Table 2.8-1.

2.8-3 *OUTPUT FORMAT*   The problem identification, $\underset{\sim}{A}$, $\underset{\sim}{b}$, $\underset{\sim}{C}$, and $\underset{\sim}{k}$ are printed out for reference. In addition the characteristic polynomial of $\underset{\sim}{F}$ is listed in both factored and unfactored form independent of which form was used to supply it.   The remainder of the output consists of elements computed in the program.

The complete $\underset{\sim}{F}$ matrix is supplied along with the observer matrices $\underset{\sim}{G}_1$ and $\underset{\sim}{G}_2$.   The output feedback coefficients $\underset{\sim}{g}$ and the compensator feedback coefficients $\underset{\sim}{h}$ are also listed [see Eqs. (2.8-11) and (2.8-12)].

2.8-4 *EXAMPLE*   Let us suppose that the basic plant is fourth order

$$\underset{\sim}{\dot{x}} = \begin{bmatrix} -2 & 1 & 0 & 0 \\ 0 & -2 & 1 & 0 \\ 0 & 0 & -1 & 1 \\ 0 & 0 & -1 & 1 \end{bmatrix} \underset{\sim}{x} + \begin{bmatrix} 0 \\ 0 \\ 0 \\ 1 \end{bmatrix} u$$

and that only x_1 and x_3 are measurable so that

$$\underset{\sim}{y} = \begin{bmatrix} 1 & 0 & 0 & 0 \\ 0 & 0 & 1 & 0 \end{bmatrix} \underset{\sim}{x}$$

and $y^* = y_1 = x_1$.   The desired $y^*(s)/r(s)$ is

---

[1]The program may be easily modified to accept any form of $\underset{\sim}{F}$ matrix.

Table 2.8-1

Input Format for LUEN Program

Card Number	Column Number	Description	Format
1	1-20	Problem identification	5A4,3I2
	21-22	N = order of the plant	
	23-24	M = dimension of output vector	
	25-26	L = order of observer	
2	1-10	a_{11}	8F10.3
	11-20	a_{12}	
	etc.	...	
N+2	1-10	b_1	8F10.3
	11-20	b_2	
	etc.	...	
N+3	1-10	c_{11}	8F10.3
	11-20	c_{12}	
	etc.	...	
N+M+3	1-10	k_1	8F10.3
	11-20	k_2	
	etc.	...	
N+M+4	1	OPTION = F (polynomial factors supplied)[1]	1A1
N+M+5	1-10	real part of the first root of observer characteristic polynomial	8F10.3
	11-20	imaginary part of ...	
N+M+L+4	1-10	real ... of the last ...	8F10.3
	11-20	imaginary part ...	

$$\frac{y^*(s)}{r(s)} = \frac{1}{s^4 + 6s^3 + 13s^2 + 13s + 7}$$

and it is easy to find by the use of STVARFDBK that K = 1.0 and
$\underline{k}$ = col(0, 1.0, 0, 1.0). Note that the state variable feedback design
requires that we feedback x_2 and x_4 but neither is available for measurement.
The program OBSERV may be used to show that ν = 2 so that a first order $\underline{F}$
matrix is needed. From STVARFDBK we may also learn that the eigenvalues of
$\underline{A}$ are all complex, hence we select $\underline{F}$ to have an eigenvalue at s = -3 so that
det($s\underline{I} - \underline{F}$) = s + 3. We are now prepared to run the program; the input deck

---

[1]The characteristic polynomial of the observer can also be entered as
polynomial coefficients by setting OPTION = P and supplying coefficients on
card N+M+5.

is shown in Table 2.8-2. The resulting output is given in Table 2.8-3. In this case we find that the observer is given by

$$\dot{z} = -3z + [-2 \quad -5]\underset{\sim}{y} + u$$

and

$$u = 1.0\{r - [1 \quad 2]\underset{\sim}{y} + z\}$$

Table 2.8-2

Input Data Deck for LUEN Example

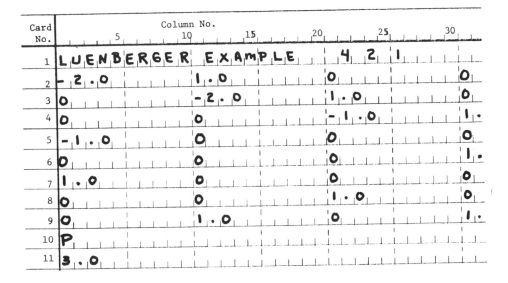

Table 2.8-3

Program Output for LUEN Example

```
LUENBERGER OBSERVER DESIGN PROGRAM
PROBLEM IDENTIFICATION:     LUENBERGER EXAMPLE

*******************************************

THE A MATRIX

  -2,0000000+00        1,0000000+00        -0,0000000        -0,00
  -0,0000000          -2,0000000+00         1,0000000+00     -0,00
  -0,0000000          -0,0000000           -1,0000000+00      1,00
  -1,0000000+00       -0,0000000           -0,0000000        -0,00

THE B MATRIX

  -0,0000000          -0,0000000           -0,0000000         1,00
```

THE C MATRIX

1,0000000+00	-0,0000000	-0,0000000	-0,00
-0,0000000	-0,0000000	1,0000000+00	-0,00

DESIRED FEEDBACK COEFFICIENTS

-0,0000000	1,0000000+00	-0,0000000	1,00

OBSERVER EIGENVALUES
   REAL PART   IMAG PART

-3,0000000-01   0,0000000

COEFFICIENTS OF OBSERVER CHARACTERISTIC POLYNOMIAL
-IN ASCENDING POWERS OF S

3,0000000-01   1,0000000+00

****

THE F MATRIX

-3,0000000-01

THE G1 MATRIX

-3,8900001+00   1,2100000+00

THE G2 MATRIX

1,0000000+00

OUTPUT FEEDBACK COEFFICIENTS

-1,7000000+00   3,0000000-01

COMPENSATOR FEEDBACK COEFFICIENTS

1,0000000+00

    2.8-5 *SUBPROGRAMS USED*

    (1)   SEMBL

    (2)   PROOT

    (3)   LINEQ

See Appendix A for a description of each of these subprograms.

```
      SUBROUTINE LUEN
C     LUENBERGER OBSERVER DESIGN PROGRAM
C     SUBROUTINES USED/ SEMBL, PROOT, LINEQ
      INTEGER R,RP,OPTION
      DATA IPP/1HP/
      DIMENSION A(10,10),B(10),C(10,10),F(10,10),H(10),AK(10),
     *  AJ(10),NAME(5),COEF(11),PHI(30,30),U(30,30),BETA(30),X(30),
     *  RR(10),RI(10)
 1000 FORMAT (5A4,3I2)
 1001 FORMAT (1H1,5X,34HLUENBERGER OBSERVER DESIGN PROGRAM/
     *  6X,23HPROBLEM IDENTIFICATION/,5X,5A4)
 1002 FORMAT (1H0,45(1H*))
 1003 FORMAT (1H0,5X,12HTHE A MATRIX/)
 1004 FORMAT (8F10.3)
 1005 FORMAT (1P6E20.7)
 1006 FORMAT (1H0,5X,12HTHE B MATRIX/)
 1007 FORMAT (1H0,5X,12HTHE C MATRIX/)
 1008 FORMAT (1H0,5X,29HDESIRED FEEDBACK COEFFICIENTS/)
 1009 FORMAT (1H0,5X,50HCOEFFICIENTS OF OBSERVER CHARACTERISTIC POLYNS
     *AL/6X,25H-IN ASCENDING POWERS OF S/)
 1011 FORMAT (1H0,5X,28HOUTPUT FEEDBACK COEFFICIENTS/)
 1012 FORMAT (1H0,5X,13HTHE G1 MATRIX/)
 1013 FORMAT (1H0,5X,12HTHE F MATRIX/)
 1014 FORMAT (1H0,5X,33HCOMPENSATOR FEEDBACK COEFFICIENTS/)
 1015 FORMAT (1H0,5X,13HTHE G2 MATRIX/)
 1016 FORMAT (1A1)
 1017 FORMAT (1H0,5X,20HOBSERVER EIGENVALUES/9X,9HREAL PART,
     *  11X,9HIMAG PART/)
      READ 1000,(NAME(I),I=1,5),N,M,R
      RP=R+1
      PRINT 1001,(NAME(I),I=1,5)
      PRINT 1002
      PRINT 1003
      DO 1 I=1,N
      READ 1004,(A(I,J),J=1,N)
    1 PRINT 1005,(A(I,J),J=1,N)
      PRINT 1006
      READ 1004,(B(I),I=1,N)
      PRINT 1005,(B(I),I=1,N)
      PRINT 1007
      DO 2 I=1,M
      READ 1004,(C(I,J),J=1,N)
    2 PRINT 1005,(C(I,J),J=1,N)
      PRINT 1008
      READ 1004,(AK(I),I=1,N)
      PRINT 1005,(AK(I),I=1,N)
      READ 1016, OPTION
      IF (OPTION.EQ.IPP) GO TO 100
      I=0
   50 I=I+1
      READ 1004,RRL,RIM
      RR(I)=-RRL
      RI(I)=RIM
      IF(RIM) 51,52,51
   51 I=I+1
      RR(I)=-RRL
      RI(I)=-RIM
   52 IF(I.LT.R) GO TO 50
      CALL SEMBL (R,RR,RI,COEF)
      GO TO 110
```

```
100 READ 1005,(COEF(I),I=1,R)
    COEF(RP)=1.0
    CALL PROOT(R,COEF,RR,RI,100)
110 PRINT 1017
    DO 120 I=1,R
120 PRINT 1005,RR(I),RI(I)
    PRINT 1009
    PRINT 1005,(COEF(I),I=1,RP)
    PRINT 1002
    DO 3 I=1,R
  3 H(I)=0.0
    H(1)=1.0
    DO 5 I=1,R
    DO 4 J=1,R
  4 F(I,J)=0.0
    F(I,1)=-COEF(RP-I)
    IF(I.EQ.R) GO TO 5
    F(I,I+1)=1.0
  5 CONTINUE
    NR=N*R
    DO 6 I=1,NR
  6 BETA(I)=0.0
    NRP=NR+1
    NRN=NR+N
    DO 7 I=1,N
  7 BETA(NR+I)=AK(I)
    DO 8 I=1,R
    II=N*(I-1)
    DO 8 J=1,R
    JJ=N*(J-1)
    DO 8 K=1,N
  8 PHI(II+K,JJ+K)=-F(I,J)
    DO 9 II=1,R
    I=(II-1)*N
    DO 9 J=1,N
    DO 9 K=1,N
  9 PHI(I+J,I+K)=PHI(I+J,I+K)+A(K,J)
    DO 10 II=1,R
    I=N*(II-1)
    DO 10 J=1,N
 10 PHI(NR+J,I+J)=H(II)
    DO 11 I=1,N
    DO 11 J=1,M
 11 PHI(NR+I,NR+J)=C(J,I)
    NRM=NR+M
    DO 12 II=1,R
    I=N*(II-1)
    J=NRM+M*(II-1)
    DO 12 K=1,M
    DO 12 L=1,N
 12 PHI(I+L,J+K)=-C(K,L)
    MM=R*(N+M)+M
    CALL LINEQ (PHI,NRN,MM,BETA,X,K,U)
    DO 13 II=1,R
    I=N*(II-1)
    AJ(II)=0.0
    DO 13 J=1,N
 13 AJ(II)=AJ(II)+B(J)*X(I+J)
    NR1=NR+1
    PRINT 1013
    DO 20 I=1,R
```

```
20 PRINT 1005,(F(I,J),J=1,R)
   PRINT 1012
   D8 14 II=1,R
   I=M*(II-1)+NRM
14 PRINT 1005,(X(J+I),J=1,M)
   PRINT 1015
   PRINT 1005,(AJ(I),I=1,R)
   PRINT 1011
   PRINT 1005,(X(J),J=NR1,NRM)
   PRINT 1014
   PRINT 1005,(H(I),I=1,R)
   RETURN
   END
```

2.9 SERIES COMPENSATION (SERCOM)

The program SERCOM is used to design a series compensation to achieve a
specified closed-loop transfer function when some of the state variables are
inaccessible.

2.9-1 *THEORY*   The work here closely follows that of Ferguson and
Rekasius[1]; it is assumed that the reader is familiar with this paper. The
plant representation will take the form

$$\dot{x}(t) = Ax(t) + bu(t) \qquad (2.9\text{-}1)$$

$$y^*(t) = c^T x(t) \qquad (2.9\text{-}2)$$

$$y(t) = Cx(t) \qquad (2.9\text{-}3)$$

Here $y^*(t)$ is the output variable to be controlled while $y(t)$ represents the
measurable output. In this method we add a series compensation to achieve
the required flexibility so that we need only the measurable output. The
dimension (order) of the compensator is equal to $\nu-1$ where ν is the observability index. The system now has the form shown in Fig. 2.9-1. The system
equations are given by

$$\dot{x} = Ax + bu \qquad (2.9\text{-}1)$$

$$\dot{z} = Dz + ew \qquad (2.9\text{-}4)$$

$$u = Kf^T z \qquad (2.9\text{-}5)$$

$$w = r - k_1^T x - k_2^T z \qquad (2.9\text{-}6)$$

Here K, k_1 and k_2 are selected to achieve a desired $(n+\nu-1)^{st}$ order closed-
loop transfer function.

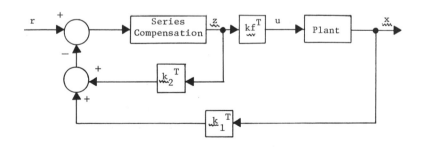

Fig. 2.9-1   Series Compensation

---

[1]See for example: J. D. Ferguson and Z. V. Rekasius, "Optimal Linear
Control Systems with Incomplete State Measurements," IEEE Transactions on
Automatic Control, Vol. AC-14, No. 2, April 1969 and J. W. Weinstein and J. L.
Melsa, "Design of Simplified Vertical Channel Landing Condition Autopilot Using
State Variables," 1970 SWIEEECO Record, Dallas, Texas, April 22-24, 1970.

We still have not solved the problem of needing the unmeasurable states. Consider a linear transformation of $\underline{z}$ of the form

$$\underline{v} = \underline{z} + \underline{T}\underline{x} \qquad (2.9\text{-}7)$$

so that

$$\dot{\underline{v}} = \dot{\underline{z}} + \underline{T}\dot{\underline{x}} = \underline{D}\underline{z} + \underline{e}w + \underline{T}\underline{A}\underline{x} + \underline{T}\underline{b}u \qquad (2.9\text{-}8)$$

Using Eq. (2.9-6) and the fact that $\underline{z} = \underline{v} - \underline{T}\underline{x}$, Eq. (2.9-8) becomes

$$\dot{\underline{v}} = (\underline{D}-\underline{e}\underline{k}_2^{T})\underline{v} + \underline{\Xi}\underline{x} + \underline{T}\underline{b}u + \underline{e}r \qquad (2.9\text{-}9)$$

where

$$\underline{\Xi} = \underline{T}\underline{A} - (\underline{D} - \underline{e}\underline{k}_2^{T})\underline{T} - \underline{e}\underline{k}_1^{T} \qquad (2.9\text{-}10)$$

If we make this transformation in Eq. (2.9-5), we obtain

$$u = K[\underline{f}^{T}\underline{v} + \underline{\xi}^{T}\underline{x}] \qquad (2.9\text{-}11)$$

where

$$\underline{\xi}^{T} = -\underline{f}^{T}\underline{T} \qquad (2.9\text{-}12)$$

We wish to select $\underline{T}$ so that $\underline{\Xi}$ and $\underline{\xi}$ only depend on the measurable output y so that

$$\underline{\Xi}\underline{x} = \underline{G}\underline{C}\underline{x} = \underline{G}\underline{y}$$

and

$$\underline{\xi}\underline{x} = \underline{g}\underline{C}\underline{x} = \underline{g}\underline{y}$$

Hence we must find $\underline{T}$, $\underline{G}$ and $\underline{g}$ to satisfy the equations:

$$\underline{T}\underline{A} - (\underline{D} - \underline{e}\underline{k}_2^{T})\underline{T} - \underline{e}\underline{k}_1^{T} = \underline{G}\underline{C} \qquad (2.9\text{-}13)$$

$$-\underline{f}^{T}\underline{T} = \underline{g}\underline{C} \qquad (2.9\text{-}14)$$

If we do this the series compensator may be written as

$$\dot{\underline{v}} = (\underline{D} - \underline{e}\underline{k}_2^{T})\underline{v} + \underline{G}\underline{y} + \underline{T}\underline{b}u + \underline{e}r$$

but $u = K[\underline{f}^{T}\underline{v} + \underline{g}^{T}\underline{y}]$ so that we have

$$\dot{\underline{v}} = (\underline{D} - \underline{e}\underline{k}_2^{T} + \underline{T}\underline{b}K\underline{f}^{T})\underline{v} + (\underline{G} + \underline{T}\underline{b}K\underline{g}^{T})\underline{y} + \underline{e}r$$

We write this result in the form

$$\dot{\underline{v}} = \overline{\underline{D}}\underline{v} + \overline{\underline{G}}\underline{y} + \underline{e}r \qquad (2.9\text{-}15)$$

where

$$\overline{\underline{G}} = \underline{G} + \underline{T}\underline{b}K\underline{g}^{T} \qquad (2.9\text{-}16)$$

$$\overline{\underline{D}} = \underline{D} - \underline{e}\underline{k}_2^{T} + \underline{T}\underline{b}K\underline{f}^{T} \qquad (2.9\text{-}17)$$

Here $\overline{\underline{G}}$ is called the major loop feedback coefficient matrix and $\underline{g}$ is called the minor loop feedback coefficient vector.

The complete design procedure is now:

(1) Find ν using OBSERV and select $\underline{D}$, $\underline{f}$, $\underline{e}$ for the compensator (arbitrarily).

(2) Select $y^*(s)/r(s)$ for augmented $(n+\nu-1)^{st}$ order system and find K, $\underline{k}_1$ and $\underline{k}_2$ as usual using STVARFDBK.

(3) Solve Eq. (2.9-13) and (2.9-14) to obtain $\underline{T}$, $\underline{G}$ and $\underline{g}$. This is the major purpose of this program.

(4) The final system is given by

$$\underline{\dot{x}} = \underline{A}\underline{x} + \underline{b}u \qquad (2.9\text{-}1)$$

$$\underline{\dot{v}} = \overline{\underline{D}}\underline{v} + \overline{\underline{G}}\underline{y} + \underline{e}r \qquad (2.9\text{-}15)$$

$$u = K[\underline{f}^T\underline{v} + \underline{g}^T\underline{y}] \qquad (2.9\text{-}18)$$

where $\overline{\underline{D}}$ and $\overline{\underline{G}}$ are defined by Eqs. (2.9-16) and (2.9-17).

2.9-2 *INPUT FORMAT* On the first data card, the problem identification, plant order, output dimension and compensator order are supplied. The next N+M+ν+4 cards supply $\underline{A}$, $\underline{b}$, $\underline{C}$, $\underline{D}$, $\underline{e}$, $\underline{f}$, $[\underline{k}_1^T \mid \underline{k}_2^T]$, and K in the usual format. See Table 2.9-1 for a detailed summary of the input format.

2.9-3 *OUTPUT FORMAT* At the beginning of the output, the problem identification, $\underline{A}$, $\underline{b}$, $\underline{C}$, $\underline{D}$, $\underline{e}$, $\underline{f}$, $[\underline{k}_1^T \mid \underline{k}_2^T]$, and K are printed for easy reference. Next the compensator system matrix $\underline{D}$ is given. This is followed by the minor_loop feedback coefficient g and the major loop feedback coefficient $\underline{G}$.

2.9-4 *EXAMPLE* Let us consider a third order plant with a single scalar output given by

$$\underline{\dot{x}}(t) = \begin{bmatrix} 0 & 1 & 0 \\ 0 & 0 & 1 \\ -6 & -11 & -6 \end{bmatrix} \underline{x}(t) + \begin{bmatrix} 0 \\ 0 \\ 1 \end{bmatrix} u(t)$$

$$y^*(t) = y(t) = [1 \quad 0 \quad 0]\underline{x}(t)$$

By the use of the OBSERV program, it is easy to determine that $\nu = 3$ so that a second order compensator is needed. The matrices $\underline{D}$, $\underline{e}$, and $\underline{f}$ may now be selected arbitrarily; the form used was

$$\underline{\dot{z}}(t) = \begin{bmatrix} 0 & 1 \\ 0 & 0 \end{bmatrix} \underline{z}(t) + \begin{bmatrix} 0 \\ 1 \end{bmatrix} w(t)$$

$$u(t) = K[1 \quad 0]\underline{z}(t)$$

We now have a fifth-order system to work with. The desired fifth order closed loop transfer function required that

$$\underline{k}_1 = \text{col}(1, 2, 1) \qquad \underline{k}_2 = \text{col}(3,4) \qquad \text{and} \qquad K = 2$$

We are now ready to use the SERCOM program to find $\overline{\underline{D}}$, g and $\overline{\underline{G}}$. The input data deck is given in Table 2.9-2. The resulting output is shown in Table 2.9-3. From this result we see that the compensator is now

71

Table 2.9-1

Input Format for SERCOM

Card Number	Column Number	Description	Format
1	1–20	Problem identification	5A4,3I2
	21–22	N = order of plant	
	23–24	M = dimension of output	
	25–26	$\nu-1$ = dimension of compensation	
2	1–10	a_{11}	8F10.3
	11–20	a_{12}	
	etc.	...	
N+1	1–10	a_{N1}	8F10.3
	11–20	a_{N2}	
	etc.	...	
N+2	1–10	b_1	8F10.3
	11–20	b_2	
	etc.	...	
N+3	1–10	c_{11}	8F10.3
	11–20	c_{12}	
	etc.	...	
N+M+2	1–10	c_{M1}	8F10.3
	11–20	c_{M2}	
	etc.	...	
N+M+3	1–10	d_{11}	8F10.3
	11–20	d_{12}	
	etc.	...	
N+M+ν+1	1–10	$d_{\nu1}$	8F10.3
	11–20	$d_{\nu2}$	
	etc.	...	
N+M+ν+2	1–10	e_1	8F10.3
	11–20	e_2	
	etc.	...	
N+M+ν+3	1–10	f_1	8F10.3
	11–20	f_2	
	etc.	...	

Table 2.9-1 (Contd.)

N+M+ν+4	1-10 11-20	k_{11} $\left.\right\}$ components of k_1 k_{12} k_{21} $\left.\right\}$ components of k_2 k_{22}	8F10.3
N+M+ν+5	1-10	K	8F10.3

$$\dot{v}(t) = \begin{bmatrix} 0 & 1 \\ -3 & -4 \end{bmatrix} v(t) + \begin{bmatrix} 2 \\ -6 \end{bmatrix} y(t) + \begin{bmatrix} 0 \\ 1 \end{bmatrix} r(t)$$

and

$$u(t) = 2.0\{ [1 \quad 0]v(t) - y(t)\}$$

2.9-5 *SUBPROGRAMS USED*

(1) LINEQ

See Appendix A for a description of this subprogram.

Table 2.9-2

Input Data Deck for SERCOM Example

Card No.	Column No.							
	5	10	15	20	25	30	35	40
1	3RD ORDER EXAMPLE				3 1 2			
2	0	1.0		0				
3	0	0		1.0				
4	-6.0	-11.0		-6.0				
5	0	0		1.0				
6	1.0	0		0				
7	0	1.0						
8	0	0						
9	0	1.0						
10	1.0							
11	1.0	2.0		1.0		3.0		4.0
12	2.0							

Table 2.9-3

Program Output for SERCOM Example

SERIES COMPENSATOR DESIGN PROGRAM
PROBLEM IDENTIFICATION-   3RD ORDER EXAMPLE

****

THE A MATRIX

-0.0000000   1.0000000+00   -0.0000000
-0.0000000   -0.0000000   1.0000000+00
-6.0000000+00   -1.1000000+01   -6.0000000+00

THE B MATRIX

-0.0000000   -0.0000000   1.0000000+00

THE C MATRIX

1.0000000+00   -0.0000000   -0.0000000

THE D MATRIX

-0.0000000   1.0000000+00
-0.0000000   -0.0000000

THE E MATRIX

-0.0000000   1.0000000+00

THE F MATRIX

1.0000000+00   -0.0000000

DESIRED FEEDBACK COEFFICIENTS

1.0000000+00   2.0000000+00   1.0000000+00

                              3.0000000+00   4.0000001+00

THE GAIN =   2.0000000+00

****

THE COMPENSATOR SYSTEM MATRIX

3.7252903-09   1.0000000+00
-3.0000000+00   -4.0000001+00

MINOR LOOP FEEDBACK COEFFICIENTS

-1.0000000+00

MAJOR LOOP FEEDBACK COEFFICIENTS

2.0000001+00
-6.0000004+00

****

75

```
      SUBROUTINE SERCOM
C        COMPENSATOR DESIGN PROGRAM
C        SUBPROGRAMS USED  LINEQ
         DIMENSION A(10,10),D(10,10),C(10,10),F(10),E(10),AK(20),
     *   PHI(30,30),BETA(30),U(30,30),X(30),NAME(5),B(10),
     *   S(10),T(10,10),P(10,10)
         INTEGER RHO
 1000 FORMAT (5A4,3I2)
 1001 FORMAT (8F10.2)
 1004 FORMAT (1H1,5X,33HSERIES COMPENSATOR DESIGN PROGRAM/
     *   6X,23HPROBLEM IDENTIFICATION=,5X,5A4)
 1005 FORMAT (1H0,45(1H*))
 1006 FORMAT (1H0,5X,12HTHE A MATRIX/)
 1007 FORMAT (1H0,5X,12HTHE B MATRIX/)
 1008 FORMAT (1H0,5X,12HTHE C MATRIX/)
 1009 FORMAT (1H0,5X,12HTHE D MATRIX/)
 1010 FORMAT (1H0,5X,12HTHE E MATRIX/)
 1011 FORMAT (1H0,5X,12HTHE F MATRIX/)
 1012 FORMAT (1H0,5X,29HDESIRED FEEDBACK COEFFICIENTS/)
 1016 FORMAT (1H0,5X,11HTHE GAIN = ,1PE20.7)
 1003 FORMAT (1P6E20.7)
 1013 FORMAT (1H0,5X,29HTHE COMPENSATOR SYSTEM MATRIX/)
 1014 FORMAT (1H0,5X,32HMINOR LOOP FEEDBACK COEFFICIENTS/)
 1018 FORMAT (1H0,5X,32HMAJOR LOOP FEEDBACK COEFFICIENTS/)
      READ 1000,(NAME(I),I=1,5),N,M,RHO
      PRINT 1004,(NAME(I),I=1,5)
      PRINT 1005
      NPRHO=N+RHO
      NN=N*(RHO+1)
      MM=RHO*(N+M)+M
      PRINT 1006
      DO 10 I=1,N
      READ 1001,(A(I,J),J=1,N)
   10 PRINT 1003,(A(I,J),J=1,N)
      PRINT 1007
      READ 1001,(B(I),I=1,N)
      PRINT 1003,(B(I),I=1,N)
      PRINT 1008
      DO 11 I=1,M
      READ 1001,(C(I,J),J=1,N)
   11 PRINT 1003,(C(I,J),J=1,N)
      PRINT 1009
      DO 13 I=1,RHO
      READ 1001,(D(I,J),J=1,RHO)
   13 PRINT 1003,(D(I,J),J=1,RHO)
      PRINT 1010
      READ 1001,(E(I),I=1,RHO)
      PRINT 1003,(E(I),I=1,RHO)
      PRINT 1011
      READ 1001,(F(I),I=1,RHO)
      PRINT 1003,(F(I),I=1,RHO)
      PRINT 1012
      READ 1001,(AK(I),I=1,NPRHO)
      PRINT 1003,(AK(I),I=1,NPRHO)
      READ 1001, GAIN
      PRINT 1016, GAIN
      PRINT 1005
      DO 20 I=1,RHO
      II=N*(I-1)
      DO 20 J=1,N
   20 BETA(II+J)=E(I)*AK(J)
```

```
      DO 30 I=1,RHO
      II=N*(I-1)
      DO 30 J=1,RHO
      JJ=N*(J-1)
      DO 30 K=1,N
   30 PHI(II+K,JJ+K)=E(I)*AK(N+J)=D(I,J)
      DO 40 II=1,RHO
      I=(II-1)*N
      DO 40 J=1,N
      DO 40 K=1,N
   40 PHI(I+J,I+K)=PHI(I+J,I+K)+A(K,J)
      NRHO=N*RHO
      DO 50 II=1,RHO
      I=N*(II-1)
      DO 50 J=1,N
   50 PHI(NRHO+J,I+J)=F(II)
      DO 60 I=1,N
      DO 60 J=1,M
   60 PHI(NRHO+I,NRHO+J)=C(J,I)
      NRHOM=NRHO+M
      DO 70 II=1,RHO
      I=N*(II-1)
      J=NRHOM+M*(II-1)
      DO 70 K=1,M
      DO 70 L=1,N
   70 PHI(I+L,J+K)=-C(K,L)
      CALL LINEQ(PHI,NN,MM,BETA,X,K,U)
      DO 80 II=1,RHO
      S(II)=0.0
      I=N*(II-1)
      DO 80 J=1,N
   80 S(II)=S(II)+X(I+J)*B(J)
      DO 110 J=1,RHO
      DO 110 I=1,RHO
  110 T(I,J)=D(I,J)=E(I)*AK(N+J)+S(I)*F(J)*GAIN
      NRHO1=NRHO+1
      DO 95 II=1,RHO
      I=M*(II-1)+NRHOM
      DO 95 J=1,M
   95 R(II,J)=X(I+J)+S(II)*X(NRHO+J)*GAIN
      PRINT 1013
      DO 90 I=1,RHO
   90 PRINT 1003,(T(I,J),J=1,RHO)
      PRINT 1014
      PRINT 1003,(X(J),J=NRHO1,NRHOM)
      PRINT 1018
      DO 130 I=1,RHO
  130 PRINT 1003,(R(I,J),J=1,M)
      PRINT 1005
      RETURN
      END
```

77

2.10 OPTIMAL CONTROL - KALMAN FILTER PROGRAM (RICATI)

The RICATI program is used to determine the matrix gains

$$K_c = R^{-1}B^T P$$

and

$$K_f = \bar{R}^{-1} C \bar{P}$$

where P and $\bar{P}$ satisfy the Riccati differential equations

$$\dot{P} = -A^T P - PA + PBR^{-1}B^T P - Q, \quad P(t_f) = P_f$$

$$\dot{\bar{P}} = A\bar{P} + \bar{P}A^T - \bar{P}C^T R^{-1}C\bar{P} + B\bar{Q}B^T, \quad \bar{P}(t_0) = \bar{P}_0$$

The origin of these equations in control problems is discussed in Schultz and Melsa[1].

At the users option, either a steady-state value of K_c or K_f or a specified number of equally spaced values of K_c or K_f on the interval $t_0 \le t \le t_f$ is printed.

2.10-1 *THEORY*   The program, in effect, makes the change of variables $\tau = t_f - t$ in the differential equation for $P(t)$ in order to put it in the same form as the equation for $\bar{P}(t)$. Under this change of variables, the equation is seen to become

$$\dot{P}(\tau) = A^T P(\tau) + P(\tau)A - P(\tau)BR^{-1}B^T P(\tau) - Q, \quad P(\tau=0) = P_f$$

The similarity of this equation to the equation for $\bar{P}(t)$ is exploited by defining, for the case where the equation for $P(t)$ is to be solved,

$$F = A^T, \quad \psi_1 = Q$$

$$E = B, \quad \psi_2 = R$$

$$D = I \quad V_0 = P_f$$

and for the case when $\bar{P}(t)$ is to be obtained

$$F = A \quad \psi_1 = \bar{Q}$$
$$E = C^T \quad \psi_2 = \bar{R}$$
$$D = B \quad V_0 = \bar{P}_0$$

Thus in either case, only the one equation

$$\dot{V} = FV + VF^T - VE\psi_2^{-1}E^T V + D\psi_1 D^T, \quad V(0) = V_0$$

need be solved. Notice that in both cases the gain K_c or K_f is given by

$$K = \psi_2^{-1}E^T V$$

The program defines F, E, D and V_0 according to the option chosen by the user

---

[1] Schultz, D. G., and J. L. Melsa, State Functions and Linear Control Systems, McGraw-Hill, New York, N.Y., 1967.

and numerically integrates the differential equation for $\underset{\sim}{V}$, according to the rule

$$\underset{\sim}{V}(t_i + EPS) = \underset{\sim}{V}(t_{i+1}) = \underset{\sim}{\dot{V}}(t_i) * EPS + \underset{\sim}{V}(t_i)$$

where EPS = .01 if only the steady-state solution is desired and EPS = .005 if values of $\underset{\sim}{K}$ on the interval $t_0 \leq t \leq t_f$ are desired.

If the user desires NPT values of $\underset{\sim}{K}$ to be printed on the interval to $t_0 \leq t \leq t_f$, $\underset{\sim}{K}$ is printed every L integration intervals, where

$$L = \frac{200 * (t_f - t_0)}{NPT}$$

and the quantity on the right is rounded to the nearest integer.

2.10-2   *INPUT FORMAT*   The input deck for the RICATI program can be divided into two sections: (1) Basic input and (2) Option input. The complete input format is summarized in Tables 2.10-1 to 2.10-3 for the user's reference. We will consider each of these sections separately. The reader should be reminded of the dimensions of the quantities involved: $\underset{\sim}{A}$(NxN), $\underset{\sim}{B}$(NxM), $\underset{\sim}{C}$(LxN).

Basic Input: The basic input contains the problem identification and the matrices $\underset{\sim}{A}$, $\underset{\sim}{B}$, and $\underset{\sim}{C}$. This input section is the same regardless of which of the two Riccati equations is to be solved. The option card, card number N+M+L+1, defines the equation to be solved and whether a steady-state or transient solution is desired. Option is set equal to C if $\underset{\sim}{K}_c$ is to be obtained and set to F if $\underset{\sim}{K}_f$ is desired. If option is left blank, the program returns to read another set of basic input. The next two variables, t_0 and t_f, are the beginning and final times for the transient response if desired. NPT is the number of points of the transient response to be printed. The user obtains the steady-state value of $\underset{\sim}{K}_c$ or $\underset{\sim}{K}_f$ by setting NPT = 0.

After solving for $\underset{\sim}{K}_c$ or $\underset{\sim}{K}_f$, the program reads another option card. Thus, the user may solve for both $\underset{\sim}{K}_c$ and $\underset{\sim}{K}_f$ with only one set of basic input. If Option is left blank, the program returns to read another set of basic input.

The option data for Option = C is shown in Table 2.10-2. Cards 1 through M define the $\underset{\sim}{R}$ matrix. Cards M + 1 through M + N define the $\underset{\sim}{Q}$ matrix. Cards M+N+1 through M+2N contain $\underset{\sim}{P}(t_f)$ only if the transient solution is desired (NPT $\neq$ 0). If NPT = 0, cards N+M+1 through M+2N are omitted.

The option data for Option = F is shown in Table 2.10-3. Cards 1 through L define the $\underset{\sim}{\bar{R}}$ matrix while cards L+1 through L+M define the $\underset{\sim}{\bar{Q}}$ matrix. Cards L+M+1 through L+M+N contain $\underset{\sim}{\bar{P}}(t_0)$ only if the transient solution is desired (NPT > 0). If NPT $\leq$ 0, cards L+M+1 through L+M+N are omitted.

2.10-3   *OUTPUT FORMAT*   The basic output of the RICATI program consists of the problem identification and input matrices. In addition, either the transient solution or the steady-state solution of the Riccati equation is printed as chosen by the user.

2.10-4   *EXAMPLE*   Consider the second-order linear system given by

$$\underset{\sim}{\dot{x}}(t) = \begin{bmatrix} -1 & 0 \\ 0 & -2 \end{bmatrix} \underset{\sim}{x}(t) + \begin{bmatrix} 1 & 0 \\ 0 & 1 \end{bmatrix} \underset{\sim}{u}(t)$$

$$\underset{\sim}{y}(t) = \begin{bmatrix} 1 & 0 \\ 0 & 2 \end{bmatrix} \underset{\sim}{x}(t)$$

TABLE 2.10-1

Basic Input Format for RICATI Program

Card Number	Column Number	Description	Format
1	1-20	Problem identification	5A4
	21-22	N = order of $\underset{\sim}{A} \leq 10$	I2
	23-24	M = number of control inputs ≤ 10	I2
	25-26	L = number of observable outputs ≤ 10	I2
2	1-10	a_{11} = Plant matrix $\underset{\sim}{A}$	8F10.3
	11-20	a_{12}	
	etc.	...	
3	1-10	a_{21}	8F10.3
	11-20	a_{22}	
	etc.	...	
N+1	1-10	b_{11} = control matrix	8F10.3
	11-20	b_{21}	
	etc.	...	
N+2	1-10	b_{12}	
	11-20	b_{22}	
	etc.	...	
N+M+1	1-10	c_{11} = output matrix	8F10.3
	11-20	c_{12}	
	etc.	...	
N+M+2	1-10	c_{21}	8F10.3
	11-20	c_{22}	
	etc.	...	
N+M+L+1	1	option	A1
	11-20	t_0	F10.3
	21-30	t_f	F10.3
	31-33	NPT	I3

TABLE 2.10-2

Option Input for Option = C

Card Number	Column Number	Description	Format
1	1-10	r_{11} = control	8F10.3
	11-20	r_{12} weighting matrix	
	etc.	...	
2	1-10	r_{21}	8F10.3
	11-20	r_{22}	
	etc.	...	
M+1	1-10	q_{11} = state	8F10.3
	11-20	q_{12} weighting matrix	
	etc.	...	
M+2	1-10	q_{21}	8F10.3
	11-20	q_{22}	
	etc.	...	
M+N+1*	1-10	$P_{11}(t_f)$	8F10.3
	11-20	$P_{12}(t_f)$	
	etc.	...	
M+N+2*	1-10	$P_{21}(t_f)$	8F10.3
	11-20	$P_{22}(t_f)$	
	etc.	...	
etc.	...	...	

*These cards included if and only if NPT > 0.

TABLE 2.10-3

Option Input for Option = F

Card Number	Column Number	Description	Format
1	1-10	$\bar{r}_{11}$ = output noise covariance matrix	8F10.3
	11-20	$\bar{r}_{12}$	
	etc.	...	
2	1-10	$\bar{r}_{21}$	8F10.3
	11-20	$\bar{r}_{22}$	
	etc.	...	
L+1	1-10	$\bar{q}_{11}$ = input noise covariance matrix	8F10.3
	11-20	$\bar{q}_{12}$	
	etc.	...	
L+2	1-10	$\bar{q}_{21}$	8F10.3
	11-20	$\bar{q}_{22}$	
	etc.	...	
L+M+1*	1-10	$\bar{P}_{21}(t_0)$	8F10.3
	11-20	$\bar{P}_{22}(t_0)$	
	etc.	...	
etc.	...	...	

*These cards are included if and only if NPT > 0

We will determine the optimal steady-state control and filter gains for

$$Q = \overline{Q} = \begin{bmatrix} 1 & 1 \\ 1 & 1 \end{bmatrix}$$

$$R = \overline{R} = \begin{bmatrix} 1 & 0 \\ 0 & 2 \end{bmatrix}$$

The complete input deck for this problem is shown in Table 2.10-4. The resulting output is given in Table 2.10-5.

TABLE 2.10-4

Input Data Deck for RICATI Example

Card No.	Column No.		
1	TRIAL PROBLEM		2 2 2
2	-1.0	0.0	
3	0.0	-2.0	
4	1.0	0.0	
5	0.0	1.0	
6	1.0	0.0	
7	0.0	2.0	
8	C		
9	1.0	0.0	
10	0.0	2.0	
11	1.0	1.0	
12	1.0	1.0	
13	F		
14	1.0	0.0	
15	0.0	2.0	
16	1.0	1.0	
17	1.0	1.0	

TABLE 2.1-5

Program Output for RICATI Program

OPTIMAL CONTROL/KALMAN FILTER PROGRAM
PROBLEM IDENTIFICATION - TRIAL PROBLEM

****

```
    THE A MATRIX
-1.00000000E 00         0.
 0.                    -2.00000000E 00

    THE B MATRIX
 1.00000000E 00         0.
 0.                     1.00000000E 00

    THE C MATRIX
 1.00000000E 00         0.
 0.                     2.00000000E 00
```

****

```
    *** CONTROL OPTION ***

    THE R MATRIX
 1.00000000E 00         0.
 0.                     2.00000000E 00

    THE Q MATRIX
 1.00000000E 00         1.00000000E 00
 1.00000000E 00         1.00000000E 00

STEADY STATE SOLUTION

    GAINS
 3.99823159E-01         2.84767617E-01
 1.42383808E-01         1.11742786E-01
```

****

```
    *** FILTER OPTION ***

    THE R MATRIX
 1.00000000E 00         0.
 0.                     2.00000000E 00

    THE Q MATRIX
 1.00000000E 00         1.00000000E 00
 1.00000000E 00         1.00000000E 00

STEADY STATE SOLUTION

    GAINS
 3.66448000E-01         2.63602536E-01
 2.63602536E-01         2.10567767E-01
```

84

2.10-5 *SUBPROGRAMS USED* The following subprograms are used by RICATI

(1) SIMEQ

See Appendix A for a description of this subprogram.

2.10-6 *PROGRAM LISTING*

```
C       RICCATI EQUATION PROGRAM (RICATI)
C       SUBROUTINES USED: SIMEQ
        DIMENSION A(10,10),B(10,10),C(10,10),R(10,10),Q(10,10),
       * K(10,10),D(10,10),NAME(5),X(10),F(10,10),E(10,10),
       *  G(10,10),H(10,10),S(10,10)
        REAL K
        INTEGER OPTION,BLANK
        DATA ICC,IFF,BLANK/1HC,1HF,1H /
   1000 FORMAT (1H1,5X,37HOPTIMAL CONTROL/KALMAN FILTER PROGRAM/)
   1001 FORMAT (5A4,3I2)
   1002 FORMAT (6X,25HPROBLEM IDENTIFICATION = ,5A4)
   1003 FORMAT (1H0,5X,13H THE A MATRIX/)
   1004 FORMAT (1H0,5X,13H THE B MATRIX/)
   1005 FORMAT(1H0,5X,13H THE C MATRIX/)
   1006 FORMAT(8F10.3)
   1007 FORMAT (A1,9X,2F10.3,I3)
   1008 FORMAT(1H0,45(1H*))
   1009 FORMAT (1H0,5X,21H*** FILTER OPTION ***/)
   1010 FORMAT (1H0,5X,22H*** CONTROL OPTION ***/)
   1011 FORMAT(6(1PE20.8))
   1012 FORMAT(1H0,5X,13H THE R MATRIX/)
   1013 FORMAT(1H0,5X,13H THE Q MATRIX/)
   1014 FORMAT(1H0,5X,19H INITIAL CONDITIONS/)
   1015 FORMAT(1H0,5X,8H TIME = ,1PE20.8/6X,5HGAINS)
   1016 FORMAT(1H0,5X,21HSTEADY STATE SOLUTION//
       *  5X,6H GAINS/)
    100 READ 1001,(NAME(I),I=1,5),N,M,L
        PRINT 1000
        PRINT 1002,(NAME(I),I=1,5)
        PRINT 1008
        PRINT 1003
        DO 110 I=1,N
        READ 1006,(A(I,J),J=1,N)
    110 PRINT 1011, (A(I,J),J=1,N)
        PRINT 1004
        DO 120 I=1,M
        READ 1006,(B(J,I),J=1,N)
    120 PRINT 1011,(B(J,I),J=1,N)
        PRINT 1005
        DO 130 I=1,L
        READ 1006,(C(I,J),J=1,N)
    130 PRINT 1011,(C(I,J),J=1,N)
    150 READ 1007,OPTION,T1,T2,NPT
        PRINT 1008
        IF(OPTION.EQ.BLANK) GO TO 100
        IF(OPTION.EQ.ICC) GO TO 300
        PRINT 1009
        NR=L
        NQ=M
        DO 230 I=1,N
        DO 210 J=1,L
    210 E(I,J)=C(J,I)
        DO 220 J=1,M
```

85

```
220 D(I,J)=R(I,J)
    DO 230 J=1,N
230 F(I,J)=A(I,J)
    GO TO 400
300 PRINT 1010
    NR=M
    NQ=N
    DO 330 I=1,N
    DO 310 J=1,M
310 E(I,J)=R(I,J)
    DO 320 J=1,L
320 D(I,J)=C(J,I)
    DO 330 J=1,N
330 F(I,J)=A(J,I)
400 PRINT 1012
    DO 410 I=1,NR
    READ 1006,(R(I,J),J=1,NR)
410 PRINT 1011,(R(I,J),J=1,NR)
    PRINT 1013
    DO 420 I=1,NQ
    READ 1006,(Q(I,J),J=1,NQ)
420 PRINT 1011,(Q(I,J),J=1,NQ)
    CALL SIMEQ(R,X,NR,H,X,IER)
    DO 440 I=1,N
    DO 440 J=1,N
    R(I,J)=0.0
    DO 440 II=1,NR
440 R(I,J)=R(I,J)+H(I,II)*E(J,II)
    IF(OPTION.EQ.ICC) GO TO 500
    DO 450 I=1,N
    DO 450 J=1,NQ
    G(I,J)=0.0
    DO 450 II=1,NQ
450 G(I,J)=G(I,J)+D(I,II)*Q(II,J)
    DO 460 I=1,N
    DO 460 J=1,N
    Q(I,J)=0.0
    DO 460 II=1,NQ
460 Q(I,J)=Q(I,J)+G(I,II)*D(J,II)
500 IF(NPT.GT.0) GO TO 530
    DO 520 I=1,N
    DO 510 J=1,N
510 G(I,J)=0.0
520 G(I,I)=1.0
    EPS=0.01
    GO TO 570
530 PRINT 1014
    DO 540 I=1,N
    READ 1006,(G(I,J),J=1,N)
540 PRINT 1011,(G(I,J),J=1,N)
    PRINT 1008
    TIME=ABS(T2-T1)
    PTS=200.0*TIME
    XX=NPT
    XX=PTS/XX
    ID=XX
    DI=ID
    YY=ABS(XX-DI)
    IF(YY.GT.0.05) ID=ID+1
    IT=PTS
    EPS=0.005
    TIME=T1
```

```
      IF(OPTION.EQ.ICC) TIME=T2
      PRINT 1015,TIME
      DO 560 I=1,NR
      DO 550 J=1,N
      K(I,J)=0.0
      DO 550 II=1,N
550   K(I,J)=K(I,J)+R(I,II)*G(II,J)
560   PRINT 1011,(K(I,J),J=1,N)
570   LC=0
      ICH=10
575   DO 580 I=1,NR
      DO 580 J=1,N
      K(I,J)=0.0
      DO 580 II=1,N
580   K(I,J)=K(I,J)+R(I,II)*G(II,J)
      DO 590 I=1,N
      DO 590 J=1,N
      H(I,J)=0.0
      DO 590 II=1,NR
590   H(I,J)=H(I,J)+E(I,II)*K(II,J)
      DO 610 I=1,N
      DO 610 J=1,N
      D(I,J)=G(I,J)
      DO 600 II=1,N
600   D(I,J)=D(I,J)+F(I,II)*G(II,J)+G(I,II)*F(J,II)-G(I,II)*H(II,J)
610   S(I,J)=G(I,J)+D(I,J)*EPS
      IF(NPT.LE.0) GO TO 640
      LC=LC+1
      IF(LC.LT.ICH) GO TO 625
      ICH=ICH+10
      ALC=LC
      T=ALC*EPS
      TIME=T1+T
      IF(OPTION.EQ.ICC) TIME=T2+T
      PRINT 1015,TIME
      DO 620 I=1,NR
620   PRINT 1011,(K(I,J),J=1,N)
      IF(LC.GE.IT) GO TO 150
625   DO 630 I=1,N
      DO 630 J=1,N
630   G(I,J)=S(I,J)
      GO TO 575
640   SUM=0.0
      DO 650 I=1,N
      DO 650 J=1,N
      SUM=SUM+ABS(D(I,J))
650   G(I,J)=S(I,J)
      IF (SUM.GT.0.0001) GO TO 575
      PRINT 1016
      DO 660 I=1,NR
660   PRINT 1011,(K(I,J),J=1,N)
      GO TO 150
      END
```

2.11 MULTI-INPUT, MULTI-OUTPUT SYSTEM DECOUPLING PROGRAM (MIMO)

The MIMO program is used to determine a feedback control law

$$\underset{\sim}{u}(t) = \underset{\sim}{G}\underset{\sim}{r}(t) + \underset{\sim}{F}\underset{\sim}{x}(t)$$

for the system Nth order system with M inputs and outputs

$$\underset{\sim}{\dot{x}}(t) = \underset{\sim}{A}\underset{\sim}{x}(t) + \underset{\sim}{B}\underset{\sim}{u}(t)$$

$$\underset{\sim}{y}(t) = \underset{\sim}{C}\underset{\sim}{x}(t), \quad \underset{\sim}{C}(M,N)$$

such that the ith input $r_i(t)$ affects only the ith output $y_i(t)$. The matrix gain $\underset{\sim}{F}$ is determined such that the closed loop poles of the ith transfer function $y_i(s)/r_i(s)$ are placed at arbitrary locations as specified by the user.

Notice that there must be exactly the same number of inputs and outputs.

2.11-1 *THEORY* The system is first transformed into "integrator decoupled" from using the algorithm described by Slivinsky[1]. At this stage, the ith input affects only the ith output with transfer function

$$y_i(s)/r_i{}^*(s) = \frac{1}{s^{m_i}}$$

The program also outputs the cancelled zeros of $y_i(s)/r_i{}^*(s)$.

At this point the program reads an option card provided by the user. If column one of the option card (card number N+2M+2) is left blank, the program returns to calculate the integrator decoupled form of the next system to be analyzed. The user may instruct the program to begin closed-loop pole-positioning for the present system by setting Option = P or F. If Option = P, the program reads the coefficients of the desired closed-loop denominator polynomial of the first subsystem $y_1(s)/r_1{}^*(s)$ from the next card. If Option = F, the program reads the desired closed-loop poles of $y_1(s)/r_1(s)$ from the next sequence of cards and returns to read the option for $y_2(s)/r_2(s)$. This will be described in more detail in the Input Format section. We will now consider the integrator decoupling and closed-loop pole placement procedures in more detail.

Let $\underset{\sim}{C}$ be represented as

$$\underset{\sim}{C} = \begin{bmatrix} \underset{\sim}{C}_1^T \\ \vdots \\ \underset{\sim}{C}_M^T \end{bmatrix}$$

where $\underset{\sim}{C}_i^T$ is the ith row of $\underset{\sim}{C}$. Define m_i to be the least integer such that

$$\underset{\sim}{C}_i^T \underset{\sim}{A}^{m_i-1} \underset{\sim}{B} \neq 0$$

and define

$$\underset{\sim}{B}^* = \begin{bmatrix} \underset{\sim}{C}_1^T \underset{\sim}{A}^{m_i-1} \underset{\sim}{B} \\ \vdots \\ \underset{\sim}{C}_M^T \underset{\sim}{A}^{m_M-1} \underset{\sim}{B} \end{bmatrix}$$

---

[1] Slivinsky, C. R. and D. G. Schultz, "State Variable Feedback Decoupling and Design of Multivariable Systems," Preprints 1969 JACC, Boulder, Colorado, 1969.

The system can be decoupled if and only if $\underset{\sim}{B}^*$ is nonsingular; i.e. if and only if $\det(\underset{\sim}{B}^*) \neq 0$.

$\underset{\sim}{B}^*$ is calculated by the program and its determinant evaluated. If $\underset{\sim}{B}^*$ if found to be singular, the message "BSTAR IS SINGULAR" is printed and the program returns to consider the next system to be decoupled. If $\underset{\sim}{B}^*$ is found to be nonsingular, the feedback control law

$$\underset{\sim}{u} = \underset{\sim}{B}^{*-1}(\underset{\sim}{v}_1 + \underset{\sim}{F}^*\underset{\sim}{x}(t))$$

is applied where

$$-\underset{\sim}{F}^* = \begin{bmatrix} \underset{\sim}{c}_1^T \underset{\sim}{A}^{m_1} \underset{\sim}{B} \\ \vdots \\ \underset{\sim}{c}_M^T \underset{\sim}{A}^{m_M} \underset{\sim}{B} \end{bmatrix}$$

This control law forces the system into integrator decoupled form. The program then proceeds to calculate, through a rather complicated algorithm, a similarity transformation $\underset{\sim}{P}$ such that with $\underset{\sim}{x}(t) = \underset{\sim}{P}\underset{\sim}{z}(t)$, the system is represented by

$$\dot{\underset{\sim}{z}} = \underset{\sim}{P}^{-1}(\underset{\sim}{A} + \underset{\sim}{B}\underset{\sim}{B}^{*-1}\underset{\sim}{F}^*)\underset{\sim}{P}\underset{\sim}{z}(t) + \underset{\sim}{P}^{-1}\underset{\sim}{B}\underset{\sim}{B}^{*-1}\underset{\sim}{v}_1(t) \quad , \quad \underset{\sim}{y}(t) = \underset{\sim}{C}\underset{\sim}{P}\underset{\sim}{z}(t)$$

where the matrices $\underset{\sim}{A}_p = \underset{\sim}{P}^{-1}(\underset{\sim}{A} + \underset{\sim}{B}\underset{\sim}{B}^{*-1}\underset{\sim}{F}^*)\underset{\sim}{P}$, $\underset{\sim}{B}_p = \underset{\sim}{P}^{-1}\underset{\sim}{B}\underset{\sim}{B}^{*-1}$, $\underset{\sim}{C}_p = \underset{\sim}{C}\underset{\sim}{P}$ have the forms

$$\underset{\sim}{A}_p = \begin{bmatrix} \underset{\sim}{A}_{11} & 0 & \cdots & 0 \\ 0 & \underset{\sim}{A}_{22} & \cdots & 0 \\ \cdots\cdots\cdots\cdots\cdots \\ 0 & \cdots & \underset{\sim}{A}_{MM} & 0 \\ \underset{\sim}{A}_{u1} & \cdots & \underset{\sim}{A}_{uM} & \underset{\sim}{A}_u \end{bmatrix}$$

$$\underset{\sim}{B}_p = \begin{bmatrix} b_1 & 0 & \cdots & 0 \\ 0 & b_2 & \cdots & 0 \\ \cdots\cdots\cdots\cdots\cdots \\ 0 & 0 & \cdots & b_M \\ b_{1u} & b_{2u} & \cdots & b_{Mu} \end{bmatrix}$$

$$\underset{\sim}{C}_p = \begin{bmatrix} \underset{\sim}{c}_{1p}^T & 0 & \cdots & 0 \\ 0 & \underset{\sim}{c}_{2p}^T & 0 & 0 \\ \cdots\cdots\cdots\cdots\cdots \\ 0 & \cdots & 0 & \underset{\sim}{c}_{Mp}^T \end{bmatrix}$$

$$\underset{\sim}{A}_{ii} = \begin{bmatrix} 0 & 1 & 0 & \cdots & 0 \\ 0 & 0 & 1 & \cdots & 0 \\ \cdots\cdots\cdots\cdots\cdots \\ 0 & \cdots & 0 & 1 \\ a_{12} & \cdots & & a_{m_i,i} \end{bmatrix}$$

$$b_i = \begin{bmatrix} 0 \\ \vdots \\ 0 \\ 1 \end{bmatrix}$$

$$c_{ip}^T = [c_{1i} \quad \cdots \quad c_{m_i,i}]$$

Thus, if z is partitioned

$$z^T = [z_1^T \quad \cdots \quad z_M^T \quad z_u^T]$$

the differential equation for z_i is

$$\dot{z}_i = A_{ii} z_i + b_i v_{1i}$$

$$y_i = c_{ip}^T z_i$$

where $v_1^T = [v_{11} \quad \cdots \quad v_{1M}]$

It is easily seen that the system has indeed been decoupled and is now represented in phase variable form. The quantities with the subscript "u" represent the uncontrollable part of the decoupled plant which has no effect on the transfer functions

$$y_i(s)/v_{1i}(s)$$

At this point, the program outputs the numerator and denominator polynomials and roots of $y_i(s)/v_{1i}(s)$. The program then reads the option card, and either calculates the closed-loop control law to yield desired $y_i(s)/v_{1i}(s)$ or returns to begin decoupling another system.

The closed loop control law is calculated by a procedure essentially the same as is used by STVARFDBK. If the desired denominator polynomial of $y_i(s)/r_i(s)$ is given in factored form, the coefficients of the polynomial are calculated by SEMBL. Then a feedback control law

$$v_{1i} = K_i[r_i(t) - k_i^T z_i(t)]$$

is determined which yields the desired closed-loop polynomial. K_i is calculated to give zero steady state error for a unit step input as in STVARFDBK.

The resulting K_i's and k_i's form matrices

$$K = \text{diag} \{K_1, \ldots, K_M\}$$

$$F = \begin{bmatrix} K_1 k_1^T & & 0 \\ & K_2 k_2^T & \\ 0 & & K_M k_M^T \end{bmatrix}$$

such that

$$\underline{v}_1(t) = \underline{K}r(t) + \underline{F}\,\underline{z}(t)$$

$\underline{K}$ is referred to as the <u>Control Gain Matrix</u> and $\underline{F}$ the <u>Feedback Gain Matrix.</u>

With the above control law, the system equation is given by

$$\dot{\underline{z}} = (\underline{A}_p + \underline{B}_p\,\underline{F})\underline{z} + \underline{B}_p\,\underline{K}\,\underline{r}$$

$$\underline{y} = \underline{C}_p\underline{z}$$

In terms of the original state variables,

$$\underline{y}(t) = \underline{C}\,\underline{x}(t)$$

$$\dot{\underline{x}}(t) = (\underline{A} + \underline{BB}^{*-1}\underline{F}^* + \underline{BB}^{*-1}\hat{\underline{F}}\hat{P}^{-1})\underline{x}(t) + \underline{BB}^{*-1}\underline{K}\underline{r}(t)$$

Thus the control law in terms of the original state variable is seen to be

$$\underline{u}(t) = \underline{B}^{*-1}\underline{K}\underline{r}(t) + \underline{B}^{*-1}(\underline{F}^* + \hat{\underline{F}}\hat{P}^{-1})\underline{x}(t)$$

The program prints $\underline{B}^{*-1}\underline{K}$ under the heading of CONTROL GAIN MATRIX and $\underline{B}^{*-1}(\underline{F}^* + \hat{\underline{F}}\hat{P}^{-1})$ under the heading of FEEDBACK GAIN MATRIX.

Thus if the above control law $\underline{u}(t)$ is applied, the resulting closed loop system will be decoupled with poles of $y_i(s)/u_i(s)$ at the locations specified by the user.

It should be noted that the design of a decoupled system in general involves two separate runs of the program. The first run is necessary to determine the order of each decoupled subsystem $y_i(s)/v_{1_i}(s)$ denoted by m_i in the program. The second run is necessary to actually compute the control law $\underline{u}(t)$ which yields the desired pole positioning as well as which decouples the system, since the number of poles of each subsystem is unknown until the first run is computed.

The authors have experienced some difficulty with numerical accuracy in this program when running problems having $\underline{A}$ matrices which are illconditioned. These problems usually manifest themselves by yielding subsystems, the sum of whose orders exceeds that of the original system. In this case, it is suggested that double precision arithmetic be used throughout the program.

2.11-2 _INPUT FORMAT_ As is shown in Table 2.11-1, the first card contains the problem identification field (columns 1-20), N and M. N is the system order, less than or equal to ten, which is punched in I2 format in columns 21-22, and M is the number of inputs and outputs, less than or equal to ten, punched in I2 format in columns 23-24. The next N cards contain the matrix $\underline{A}$, entered as usual by rows. The next M cards contain the matrix B, which is read as $\underline{B}^T$; i.e. the first card contains the first column of B, etc. The next M cards contain the rows of $\underline{C}$. All the above matrices are entered in 8F10.3 format.

The next card is the OPTION card. If this card is left blank, the program obtains the phase variable decoupled form of the system and returns to begin another problem. If OPTION, in A1 format, is either P or F, the program determines the closed-loop control law $\underline{u}(t)$ as described above. If OPTION = P, the next card is read to determine the M_1 desired coefficients of the closed loop polynomial $y_1(s)/r_1(s)$. The coefficient of s^{m_1} is assumed to be unity. If OPTION = F, the m_1 roots of $y_1(s)/r_1(s)$ are read from the next set of cards. As usual, left-half-plane poles are entered with positive real parts and only one root of a complex conjugate pair is entered. The roots are read one per card. The program then returns to read OPTION for the second subsystem and either the polynomial coefficients or desired poles for the second subsystem, etc.

 2.11-3 *OUTPUT FORMAT* The problem identification and the matrices A, B, c are printed first for easy reference.

 The decoupled phase variable representation of each subsystem is then printed. The denominator polynomial is printed first, then the numerator polynomial is printed, both in ascending powers of s. The numerator polynomial is then factored and the zeros are listed.

 In the closed-loop design mode, the desired closed-loop polynomial and pole locations for each subsystem are repeated. Finally, the Control Gain and Feedback Gain Matrices are printed for the original state variable description of the system.

 2.11-4 *EXAMPLE* A combined analysis-design problem for the third order system given by

$$\dot{\underset{\sim}{x}}(t) = \begin{bmatrix} -1 & 0 & 0 \\ 0 & -2 & 0 \\ 0 & 0 & -3 \end{bmatrix} \underset{\sim}{x}(t) + \begin{bmatrix} 1 & 0 \\ 0 & 1 \\ 0 & 1 \end{bmatrix} \underset{\sim}{u}(t)$$

$$\underset{\sim}{y}(t) = \begin{bmatrix} 1 & 1 & 0 \\ 0 & 1 & 1 \end{bmatrix} \underset{\sim}{x}(t)$$

is considered. The subsystem orders have been previously determined to be $m_1 = 1$, $m_2 = 2$. The desired closed loop pole location for subsystem one is $s = -3$, and $s = -1 \pm j1$ for subsystem two.

 The complete input deck for this example is shown in Table 2.11-3 and the resulting output in Table 2.11-4.

 2.11-5 *SUBPROGRAMS USED* The following subprograms are used by MIMO

  (1) HERMIT
  (2) MULT
  (3) DET
  (4) PROOT
  (5) SEMBL
  (6) SIMEQ
  (7) CHREQA

See Appendix A for a description of these subprograms.

TABLE 2.11-1

Input Format for MIMO

Card Number	Column Number	Description	Format
1	1-20	Problem Identification	5A4
	21-22	N = system order	I2
	23-24	M = number of inputs and outputs	I2
2	1-10	a_{11}	8F10.3
	11-20	a_{12}	
	⋮	⋮	
N+1	1-10	a_{N1}	8F10.3
	11-20	a_{N2}	
	etc.	⋮	
N+2	1-10	b_{11}	8F10.3
	11-20	b_{21}	
	etc.	⋮	
N+3	1-10	b_{12}	8F10.3
	11-20	b_{22}	
	etc.	⋮	
N+M+1	1-10	b_{1M}	8F10.3
	11-20	b_{2M}	
	etc.	⋮	
N+M+2	1-10	c_{11}	8F10.3
	11-20	c_{12}	
	etc.	⋮	
N+2M+1	1-10	c_{M1}	8F10.3
	11-20	c_{M2}	
	etc.	⋮	

TABLE 2.11-1 (Continued)

N+2M+2	1	OPTION $\begin{cases} \text{Blank = analysis only} \\ \text{P = Polynomial input} \\ \text{F = Factored input} \end{cases}$	A1
N+2M+2	Closed loop design data if OPTION = P or OPTION = F See TABLE 2.11-2.		

TABLE 2.11-2

Closed Loop Design Data for OPTION = P or F

Card Number	Column Number	Description	Format
0	1	OPTION = P*	A1
1	1-10 11-20 etc.	α_1 α_2 $\vdots$	8F10.3

Card Number	Column Number	Description	Format
0	1	OPTION = F	A1
1	1-10 11-20	real part of first root imaginary part of first root	8F10.3
2	1-10 11-20	real part of 2nd root imaginary part of 2nd root	8F10.3
etc.	etc.		

*The desired closed loop denominator polynomial for the subsystem
$y_i/r_i(s)$ is of the form

$$P_I(s) = s^{m_i} + \alpha_{m_i} s^{m_i-1} + \ldots + \alpha_2 s + \alpha_1$$

TABLE 2.11-3

Input Deck for MIMO Example

Card No.	Column No. (1 ... 5 ... 10 ... 15 ... 20 ... 25)
1	MIMO EXAMPLE ONE 3 2
2	-1.0 0.0 0.0
3	0.0 -2.0 0.0
4	0.0 0.0 -3.0
5	1.0 0.0 0.0
6	0.0 1.0 1.0
7	1.0 1.0 0.0
8	0.0 1.0 1.0
9	P
10	3.0
11	F
12	1.0 1.0

Table 2.11-4

Program Output for MIMO Program

MULTI-INPUT, MULTI-OUTPUT PROGRAM

PROBLEM IDENTIFICATION -   MIMO EXAMPLE ONE

****

THE A MATRIX

-1.00000+00	-0.00000	-0.00000
-0.00000	-2.00000+00	-0.00000
-0.00000	-0.00000	-3.00000+00

THE B MATRIX

1.00000+00	-0.00000	-0.00000
-0.00000	1.00000+00	1.00000+00

THE C MATRIX

1.00000+00	1.00000+00	-0.00000
-0.00000	1.00000+00	1.00000+00

****

DECOUPLED PHASE VARIABLE REPRESENTATION

*** SUBSYSTEM   1

95

DENOMINATOR POLYNOMIAL - IN ASCENDING POWERS OF S

-0.00000   1.00000+00

   NUMERATOR POLYNOMIAL - IN ASCENDING POWERS OF S

   1.00000+00

*** SUBSYSTEM 2

   DENOMINATOR POLYNOMIAL - IN ASCENDING POWERS OF S

-0.00000   2.50000+00   1.00000+00

   NUMERATOR POLYNOMIAL - IN ASCENDING POWERS OF S

   2.50000+00   1.00000+00

   ZEROS OF NUMERATOR POLYNOMIAL
                  REAL PART   IMAGINARY PART
                 -2.5000000+00   0.0000000

.**

   CLOSED-LOOP CALCULATIONS

.**

*** SUBSYSTEM 1

   CLOSED-LOOP POLYNOMIAL - IN ASCENDING POWERS OF S

   3.00000+00   1.00000+00

   CLOSED-LOOP POLES   REAL PART   IMAGINARY PART
                               -3.0000000+00   0.0000000

*** SUBSYSTEM 2

   CLOSED-LOOP POLYNOMIAL - IN ASCENDING POWERS OF S

   2.00000+00   2.00000+00   1.00000+00

   CLOSED-LOOP POLES   REAL PART   IMAGINARY PART
                               -1.0000000+00   1.0000000+00
                               -1.0000000+00   -1.0000000+00

.**

   ORIGINAL STATE VARIABLES

   FEEDBACK GAIN MATRIX

-2.00000+00   1.00000+00   -5.00000+00
 0.00000   -2.00000+00   5.00000+00

   CONTROL GAIN MATRIX

   3.00000+00   -4.00000-01
   0.00000   4.00000-01

.**

96

```
C     MULTI-INPUT MULTI-OUTPUT PROGRAM (MIMO)
C     SUBPROGRAMS USED: HERMIT, MULT, DET, PROOT, SEMBL, SIMEQ,
      DIMENSION A(10,10),B(10,10),C(10,10),NAME(5),AHAT(10,10),
     *  BHAT(10,10),CHAT(10,10),CI(10,10),CIB(10,10),AKB(10,10),
     *  ASTAR(10,10),BSTAR(10,10),Y(11),X(11),QT(10,10),
     *  QI(10,10),BT(10,10),FSTAR(10,10),AID(10,10),BID(10,10),
     *  FORQ(20,10),Q(10,10),QIN (20,10),QTT(10,10),BASIS(10,10),
     *  BIG(10,10),GAIN(10),P(10,10),PIN(10,10),COEF(11),
     *  AP(10,10),BP(10,10),CP(10,10),XX(10),YY(10)
      DIMENSION JCOLM(10),NSUBI(10),IRANK(10)
      INTEGER D(10),RANK1,RANK2,BLANK,POLY,OPTION
      DATA BLANK,POLY/1H ,1HP/
 1000 FORMAT (1H0,45(1H*))
 1001 FORMAT (5A4,2I2)
 1002 FORMAT (8F10.3)
 1003 FORMAT (6X,25HPROBLEM IDENTIFICATION -  ,5X,5A4)
 1004 FORMAT (1H0,5X,12HTHE A MATRIX/)
 1005 FORMAT (1H0,5X,12HTHE B MATRIX/)
 1006 FORMAT (1H0,5X,12HTHE C MATRIX/)
 1007 FORMAT (1P8E15.5)
 1008 FORMAT (1H1,5X,33HMULTI-INPUT, MULTI-OUTPUT PROGRAM   /)
 1009 FORMAT(1H0,5X,23HDENOMINATOR POLYNOMIAL ,
     *  26H- IN ASCENDING POWERS OF S//1P8E15.5)
 1010 FORMAT(1H0,5X,21HNUMERATOR POLYNOMIAL ,
     *  26H- IN ASCENDING POWERS OF S//1P8E15.5)
 1011 FORMAT(1H0,5X,16HDECOUPLED PHASE ,
     *  23HVARIABLE REPRESENTATION)
 1012 FORMAT(A1)
 1013 FORMAT(1H0,15H**** SUBSYSTEM ,I2)
 1014 FORMAT (1H0,5X,22HCLOSED-LOOP POLYNOMIAL,
     *  27H - IN ASCENDING POWERS OF S/)
 1015 FORMAT (1H0,5X,17HCLOSED-LOOP POLES ,10X,
     *  9HREAL PART, 10X,14HIMAGINARY PART)
 1016 FORMAT(25X,1P2E20.7)
 1017 FORMAT(1H0,5X,19HZEROS OF NUMERATOR ,
     *  10HPOLYNOMIAL/16X,9HREAL PART,10X,14HIMAGINARY PART)
 1019 FORMAT(8X,1P2E20.7)
 1020 FORMAT(1H0,5X,24HORIGINAL STATE VARIABLES/
     *  1H0,5X,20HFEEDBACK GAIN MATRIX/)
 1021 FORMAT(1H0,5X,19HCONTROL GAIN MATRIX/)
 1022 FORMAT(1H0,5H**** ,24HUNCONTROLLABLE SUBSYSTEM)
 1023 FORMAT(1H0,5X,23HDENOMINATOR POLYNOMIAL ,
     *  26H- IN ASCENDING POWERS OF S/)
 1024 FORMAT(1H0,5X,21HUNCONTROLLABLE POLES ,
     *  6X,9HREAL PART,10X,14HIMAGINARY PART)
 1025 FORMAT(1H0,5X,24HCLOSED-LOOP CALCULATIONS)
 1992 FORMAT (/8X,17HBSTAR IS SINGULAR/)
   50 READ(5,1001,END=2000) (NAME(I),I=1,5),N,M
      PRINT 1008
      PRINT 1003,(NAME(I),I=1,5)
      PRINT 1000
      PRINT 1004
      DO 100 I=1,N
      READ 1002,(A(I,J),J=1,N)
  100 PRINT 1007,(A(I,J),J=1,N)
      PRINT 1005
      DO 101 I=1,M
      READ 1002,(B(J,I),J=1,N)
  101 PRINT 1007,(B(J,I),J=1,N)
      PRINT 1006
```

```
      DO 102 I=1,M
      READ 1002,(C(I,J),J=1,N)
  102 PRINT 1007, (C(I,J),J=1,N)
      PRINT 1000
      DO 103 I=1,N
      Y(I)=0.0
      DO 103 J=1,N
  103 QI(I,J)=0.0
      DO 104 I=1,N
  104 QI(I,I)=1.0
C     COMPUTE BSTAR, ASTAR
      ZAID=1.0E-8
      DO 150 I=1,M
      D(I)=0
      DO 105 J=1,N
  105 CI(1,J)=C(I,J)
      CALL MULT (CI,B,CIB,1,N,M)
      DSUM=0.0
      DO 106 J=1,M
  106 DSUM=DSUM+ ABS(CIB(1,J))
      IF (DSUM.LT.ZAID) GO TO 109
      DO 107 J=1,M
  107 BSTAR(I,J)=CIB(1,J)
      CALL MULT(CI,A,CIB,1,N,N)
      DO 108 J=1,N
  108 ASTAR(I,J) = CIB(1,J)
      GO TO 150
  109 NMIN=N-1
      DO 110 K=1,N
  110 AKB(1,K)=CI(1,K)
      DO 145 K=1,NMIN
      CALL MULT(AKB,A,AKB,1,N,N)
      CALL MULT(AKB,B,CIB,1,N,M)
      DSUM=0.0
      DO 111 J=1,M
  111 DSUM=DSUM+ ABS(CIB(1,J))
      IF(DSUM.LT.ZAID.AND.K.NE.NMIN) GO TO 145
      IF(DSUM.LT.ZAID.AND.K.EQ.NMIN) GO TO 130
      DO 112 J=1,M
  112 BSTAR(I,J)=CIB(1,J)
      CALL MULT(CI,A,AKB,1,N,N)
      DO 120 L=1,K
      CALL MULT(AKB,A,AKB,1,N,N)
  120 CONTINUE
      DO 125 J=1,N
  125 ASTAR(I,J)=AKB(1,J)
      D(I)=K
      GO TO 150
  130 DO 135 J=1,M
  135 BSTAR(I,J)=0.0
      DO 140 J=1,N
  140 ASTAR(I,J)=0.0
      D(I)=NMIN
      GO TO 150
  145 CONTINUE
  150 CONTINUE
      IF( ABS(DET(BSTAR,M)).GE.ZAID) GO TO 160
      PRINT 1992
      STOP
  160 CALL SIMEQ(BSTAR,Y,M,BT,X,IER)
      CALL MULT(BT,ASTAR,FSTAR,M,M,N)
      DO 175 I=1,M
      DO 175 J=1,N
```

```
175 FSTAR(I,J)=-FSTAR(I,J)
    CALL MULT(B,FSTAR,CIB,N,M,N)
    DO 185 I=1,N
    DO 185 J=1,N
    AID(I,J)=A(I,J)+CIB(I,J)
    IF( ABS(AID(I,J)).LT.ZAID)AID(I,J)=0.0
185 CONTINUE
    CALL MULT(B,BT,BID,N,M,M)
    DO 200 I=1,N
    DO 200 J=1,M
    BIG(I,J)=BID(I,J)
200 CI(I,J)=BID(I,J)
    LL=N-1
    DO 205 I=1,LL
    CALL MULT(AID,CI,CI,N,N,M)
    LA =I*M+1
    LB=(I+1)*M
    DO 205 J=LA,LB
    LC=J-LA+1
    DO 205 K=1,N
205 BIG(K,J)=CI(K,LC)
    LIQ=0
    JSUM=0
    DO 350 IQ=1,M
    I=1
    J=1
    NM=N*M
    DO 225 K=1,NM
    IF(K.NE.(J-1)*M+IQ) GO TO 215
210 IF(K.LT.J*M) GO TO 225
    J=J+1
    GO TO 225
215 DO 220 L=1,N
220 FORQ(I,L)=BIG(L,K)
    I=I+1
    GO TO 210
225 CONTINUE
    ICOL=N*(M-1)
    CALL HERMIT(FORQ,ICOL,N,IRANK(IQ))
    IJK=IRANK(IQ)
    DO 235 I=1,IJK
    DO 230 J=1,N
    IF(FORQ(I,J).LT..99.OR.FORQ(I,J).GT.1.01) GO TO 230
    JCOLM(I)=J
    GO TO 235
230 CONTINUE
235 CONTINUE
    NBASIS=N-IJK
    NSUBI(IQ)=NBASIS
    J=1
    DO 265 I=1,NBASIS
240 DO 245 K=1,IJK
    IF(J.EQ.JCOLM(K)) GO TO 255
245 CONTINUE
    DO 250 L=1,IJK
250 CHAT(I,L)=-FORQ(L,J)
    GO TO 260
255 J=J+1
    GO TO 240
260 J=J+1
265 CONTINUE
    DO 275 J=1,IJK
    JTEMP=JCOLM(J)
```

```
      DO 270 K=1,NBASIS
270 BASIS(K,JTEMP)=CHAT(K,J)
275 CONTINUE
      JJ=1
      DO 300 J=1,NBASIS
280 DO 285 K=1,IJK
      IF(JJ.EQ.JCOLM(K)) GO TO 295
285 CONTINUE
      DO 290 L=1,NBASIS
290 BASIS(L,JJ)=QI(L,J)
      JJ=JJ+1
      GO TO 300
295 JJ=JJ+1
      GO TO 280
300 CONTINUE
      JSUM=JSUM+NBASIS
      DO 305 ICOL=1,N
      QT(1,ICOL)=C(IQ,ICOL)
305 CI(1,ICOL)=C(IQ,ICOL)
      IF(D(IQ).EQ.0) GO TO 315
      LLL=D(IQ)
      DO 310 JJ=1,LLL
      CALL MULT(CI,AID,CI,1,N,N)
      DO 310 ICOL=1,N
310 QT(JJ+1,ICOL)=CI(1,ICOL)
315 IF(NBASIS.EQ.D(IQ)+1) GO TO 340
      RANK1=D(IQ)+1
      DO 335 II=1,NBASIS
      DO 320 IROW=1,RANK1
      DO 320 ICOL=1,N
320 QIN(IROW,ICOL)=QT(IROW,ICOL)
      DO 325 ICOL=1,N
325 QIN(RANK1+1,ICOL)=BASIS(II,ICOL)
      IRAN=RANK1+1
      CALL HERMIT(QIN,IRAN,N,RANK2)
      IF(RANK1.EQ.RANK2) GO TO 335
      DO 330 ICOL =1,N
330 QT(IRAN,ICOL)=BASIS(II,ICOL)
      RANK1=IRAN
      IF(RANK1.EQ.NBASIS) GO TO 340
335 CONTINUE
340 IA=LIQ+1
      IB=LIQ+NBASIS
      IR=0
      DO 345 IROW=IA,IB
      IR=IR+1
      DO 345 ICOL=1,N
345 Q(IROW,ICOL)=QT(IR,ICOL)
      LIQ=LIQ+NBASIS
350 CONTINUE
      IF(JSUM.EQ.N) GO TO 422
      II=JSUM
      DO 420 III=1,N
      DO 408 I=1,JSUM
      DO 408 J=1,N
408 FORQ(I,J)=Q(I,J)
      DO 412 J=1,N
412 FORQ(JSUM+1,J)=QI(III,J)
      CALL HERMIT(FORQ,JSUM+1,N,KRANK)
      IF(KRANK.EQ.JSUM) GO TO 420
      II=II+1
      DO 416 J=1,N
416 Q(II,J)=QI(III,J)
```

100

```
      JSUM=JSUM+1
      IF(JSUM.EQ.N) GO TO 422
420   CONTINUE
422   CONTINUE
      CALL SIMEQ(Q,Y,N,QI,X,IER)
      CALL MULT(AID,QI,AHAT,N,N,N)
      CALL MULT(Q,AHAT,AHAT,N,N,N)
      CALL MULT(Q,BID,BHAT,N,N,M)
      CALL MULT(C,QI,CHAT,M,N,N)
      DO 460 I=1,N
      DO 460 J=1,N
      P(I,J)=0.0
460   PIN(I,J)=0.0
      NRUN=0
      DO 480 I=1,M
      NSUB=NSUBI(I)
      LOW=NRUN
      NRUN=NRUN+NSUB
      DO 474 II=1,NSUB
      JJ=LOW+II
474   QT(II,NSUB)=BHAT(JJ,I)
      IF(NSUB.EQ.1) GO TO 476
      DO 470 II=1,NSUB
      DO 470 J=1,NSUB
      JJ=LOW+II
      KK=LOW+J
470   AKB(II,J)=AHAT(JJ,KK)
      CALL CHREQA(AKB,NSUB,COEF)
      DO 475 JJ=2,NSUB
      KK=NSUB-JJ+1
      LL=KK+1
      DO 475 II=1,NSUB
      JJJ=LOW+II
      QT(II,KK)=COEF(LL)*BHAT(JJJ,I)
      DO 475 MM=1,NSUB
475   QT(II,KK)=QT(II,KK)+AKB(II,MM)*QT(MM,LL)
476   CALL SIMEQ(QT,Y,NSUB,QTT,X,IER)
      DO 477 J=1,NSUB
      LPJ=LOW+J
      DO 477 K=1,NSUB
      LPK=LOW+K
      P(LPJ,LPK)=QT(J,K)
477   PIN(LPJ,LPK)=QTT(J,K)
480   CONTINUE
      IF(NRUN.EQ.N) GO TO 502
      NST=NRUN+1
      DO 501 I=NST,N
      PIN(I,I)=1.0
501   P(I,I)=1.0
502   CALL MULT(PIN,AHAT,AP,N,N,N)
      CALL MULT(AP,P,AP,N,N,N)
      CALL MULT(PIN,BHAT,BP,N,N,M)
      CALL MULT(CHAT,P,CP,M,N,N)
      PRINT 1011
      KOUNT=0
      DO 503 I=1,M
      PRINT 1013,I
      NSUB=NSUBI(I)
      NP=NSUB+1
      K=KOUNT+NSUB
      DO 504 J=1,NSUB
      KPJ=KOUNT+J
      X(J)=-AP(K,KPJ)
```

101

```
504 Y(J)=CP(I,KPJ)
    X(NP)=1.0
    DO 505 J=1,NSUB
    NPM=N-J
    IF(ABS(Y(NPM)).GE.ZAID) GO TO 506
505 CONTINUE
506 NUM=NSUB-J
    NNN=NUM+1
    PRINT 1009,(X(J),J=1,NP)
    PRINT 1010,(Y(J),J=1,NNN)
    IF(NUM.EQ.0)GO TO 503
    CALL PROOT(NUM,Y,XX,YY,1)
    PRINT 1017
    DO 507 J=1,NUM
507 PRINT 1019,XX(J),YY(J)
503 KOUNT=KOUNT+NSUB
    IF(KOUNT.GE.N) GO TO 520
    PRINT 1022
    K=N-KOUNT
    DO 510 I=1,K
    DO 510 J=1,K
    KPI=KOUNT+I
    KPJ=KOUNT+J
510 AKB(I,J)=AP(KPI,KPJ)
    CALL CHREQA(AKB,K,X)
    KP=K+1
    PRINT 1023
    PRINT 1007,(X(I),I=1,KP)
    CALL PROOT(K,X,XX,YY,1)
    PRINT 1024
    DO 515 I=1,K
515 PRINT 1016,XX(I),YY(I)
520 KOUNT=0
    NFLAG=0
    DO 516 I=1,N
    DO 516 J=1,N
516 QT(I,J)=0.0
    DO 650 II=1,M
    READ (5,1012) OPTION
    NSUB=NSUBI(II)
    NP=NSUB+1
    IF(OPTION.EQ.BLANK) GO TO 50
    IF(NFLAG.NE.0)GO TO 575
    PRINT 1000
    PRINT 1025
    PRINT 1000
    NFLAG=100
575 IF(OPTION.EQ.POLY) GO TO 630
    K=0
600 K=K+1
    READ 1002,RR,RI
    X(K)=-RR
    Y(K)=RI
    IF( ABS(RI).LT.ZAID) GO TO 610
    K=K+1
    X(K)=-RR
    Y(K)=-RI
610 IF(K.LT.NSUB) GO TO 600
    CALL SEMBL(NSUB,X,Y,COEF)
    GO TO 620
630 READ 1002,(COEF(I),I=1,NP)
    COEF(NP)=1.0
```

```
      CALL PROOT(NSUB,COEF,X,Y,1)
  620 PRINT 1013,II
      PRINT 1014
      PRINT 1007,(COEF(I),I=1,NP)
      PRINT 1015
      DO 635 I=1,NSUB
  635 PRINT 1016,X(I),Y(I)
      GAIN(II)=COEF(1)/CP(II,KOUNT+1)
      K=KOUNT+NSUB
      DO 640 I=1,NSUB
      KPI=KOUNT+I
  640 QT(II,KPI)=-(COEF(I)+AP(K,KPI))/GAIN(II)
  650 KOUNT=K
      DO 800 J=1,M
      G=GAIN(J)
      DO 800 I=1,M
  800 BT(I,J)=G*BT(I,J)
      CALL MULT(BT,QT,QTT,M,M,N)
      CALL MULT(QTT,PIN,CI,M,N,N)
      CALL MULT(CI,Q,QTT,M,N,N)
      DO 805 I=1,M
      DO 805 J=1,N
  805 CI(I,J)=FSTAR(I,J)+QTT(I,J)
      PRINT 1000
      PRINT 1020
      DO 810 I=1,M
  810 PRINT 1007,(CI(I,J),J=1,N)
      PRINT 1021
      DO 820 I=1,M
  820 PRINT 1007,(BT(I,J),J=1,M)
      PRINT 1000
      GO TO 50
 2000 STOP
      END
```

3 TRANSFER FUNCTION PROGRAMS

3.1 INTRODUCTION

We will examine in this chapter three programs which may be used to solve certain computation problems associated with transfer functions. These programs may be used to obtain frequency response and root locus plots and to form partial fraction expansions of rational functions.

The FREQUENCY RESPONSE (FRESP) program discussed in Sec. 3.2, allows one to obtain and plot the frequency response of a rational transfer function over a specified range of frequencies. Both rectangular Bode plots as well as a polar Nyquist plot can be obtained as an output of the program as desired.

Root locus plots are obtained from the ROOT LOCUS (RTLOC) program which is discussed in Sec. 3.3. In a sense this program complements the SENSIT program of Sec. 2.5. In addition to the normal root locus plot, one may choose to examine only a small rectangular section of the root locus in more detail.

As the transfer function version of the time response programs GTRESP and especially RTRESP, the PARTIAL FRACTION EXPANSION (PRFEXP) program is used to obtain the partial fraction expansion of a rational transfer function. This program, which is discussed in Sec. 3.4, may be used on transfer functions with multiple real roots although complex roots are assumed to be simple.

3.2 FREQUENCY RESPONSE (FRESP)

The FRESP program is used for determining the frequency response of a rational transfer function G(s) of the form

$$G(s) = K \frac{A(s)}{B(s)}$$

where

$$A(s) = a_1 + a_2 s + \ldots + a_M s^{M-1} + s^M$$

$$B(s) = b_1 + b_2 s + \ldots + b_N s^{N-1} + s^N$$

The frequency response may be graphically displayed in the form of Bode and/or Nyquist diagrams.

3.2-1 *THEORY*   The complex number $G(j\omega)$ can be computed for specific values of ω by simple complex algebra by substitution of $j\omega$ for s. If we write $G(j\omega)$ in rectangular form as

$$G(j\omega) = R(\omega) + jX(\omega)$$

where $R(\omega)$ and $X(\omega)$ are real values, then the magnitude and phase of $G(j\omega)$ are given by

$$|G(j\omega)| = [R^2(\omega) + X^2(\omega)]^{1/2}$$

$$\arg G(j\omega) = \arctan \frac{X(\omega)}{R(\omega)}$$

The program will linearly or logarithmically interpolate frequency values between two extreme values or discrete values of frequency may be supplied.

3.2-2 *INPUT FORMAT*   The input deck for the FRESP program can be divided into three separate sections: (1) Basic input, (2) Option card and (3) Values of ω. The complete input format is summarized in Table 3.2-1. Let us consider each of these input sections individually.

Basic Input: The basic input for this program consists of problem identification and the discription of the rational function G(s). Both the numerator and denominator polynomial may be described as either polynomial coefficients (KEY = P) or polynomial factors (KEY = F). Note that if the coefficient mode is selected, the coefficient of the highest order term in both polynomial is unity. For simplicity, the polynomial A(s) shown in P mode input and B(s) in F mode for Table 3.2-1.

Option Card: The option card is used to select which mode of frequency-value input is to be used and which graphical displays are to be used. The value of KNOW (Column 26) is used to select the type of frequency range to be used. If KNOW = 0, extreme values of frequency are read from two ten column fields and logarithmic interpolation is used to generate values between these extremes; the total number of frequencies to be used is indicated by NOMEG. For KNOW = 2, exactly the same procedure is used except linear interpolation of frequency values is used rather than logarithmic interpolation.

When KNOW = 1, the program reads NOMEG values of frequency from the cards which follow the option card. This is the only case in which third section of the input deck is used. Note that NOMEG must be less than or equal to 500.

The variables NBODE and NYQST are used to indicate whether a Bode and/or Nyquist plot of the frequency response. If KNOW = 0, both graphical forms are available; if KNOW = 1, neither graphical output may be used and if KNOW = 2 only the Nyquist plot may be obtained. Table 3.2-2 shows the relation of KNOW and the available graphical outputs.

Table 3.2-1

Input Format for FRESP Program

Card Number	Column Number	Description	Format
1	1-20	Problem identification	5A4
2	1-10	K = gain	8E10.0
3	1	KEY = P (coefficient supplied)[1]	A1,I2
	2-3	M = order of A(s) polynomial	
4	1-10	a_1 = coefficients of A(s)	8E10.0
	11-20	a_2	
	etc.	...	
5	1	KEY = F (polynomial factors supplied)[1]	A1, I2
	2-3	N = order of B(s) polynomial	
6	1-10	real part of the first root of B(s)	8E10.0
	11-20	imaginary part of the first root of B(s)	
7	1-10	real part of the second root of B(s)	8E10.0
	11-20	imaginary part of the second root of B(s)	
N+6	1-10	WMIN = Minimum frequency value	2E10.0, 4I3
	11-20	WMAX = Maximum frequency value	
	21-23	NOMEG = number of frequency values to be used, NOMEG $\leq$ 500	
	24-25	NOT USED	
	26	KNOW = 0 logarithmic interpolation	
		= 1 discrete value supplied	
		= 2 linear interpolation	
	27-28	NOT USED	
	29	NBODE = 0 Bode plot desired	
		= 1 Bode plot not desired	
	30-31	NOT USED	
	32	NYQST = 0 Nyquist plot desired	
		= 1 Nyquist plot not desired	
N+7	1-10	WFREQ(1) } This card is only used	8E10.0
	11-20	WFREQ(2) } if KNOW = 1.	
etc.	...	...	

[1]Both A(s) and B(s) may be supplied in either P or F form.

Table 3.2-2

Values of KNOW and Available Graphical Output

KNOW	Bode Plot	Nyquist Plot
0	Yes	Yes
1	No	No
2	No	Yes

Values of Frequency: This section of the input data deck is only used if KNOW = 1 indicating that discrete values of frequency are to be read in. The number of values of frequency to be read is NOMEG and these values are read, eight per card, from the next cards following the option card.

3.2-3 *OUTPUT FORMAT* The output of the FRESP program begins with the problem identification contained on the first input card. Next the coefficients and roots of the polynomials A(s) and B(s) and the gain K are listed for reference. The actual frequency response information is listed in tabular form with the following items:

(1) RADIAN FREQUENCY - the values of ω either read in (KNOW = 1) or calculated by the program (KNOW = 0 or = 2)
(2) REAL PART - $R(\omega)$, the real part of $G(j\omega)$
(3) IMAG. PART - $X(\omega)$, the imaginary part of $G(j\omega)$
(4) MAGNITUDE - $|G(j\omega)| = [R(\omega)^2 + X(\omega)^2]^{1/2}$, the magnitude of $G(j\omega)$
(5) PHASE (RAD) - arg $[G(j\omega)] = \tan^{-1}[X(\omega)/R(\omega)]$, the phase angle in radians
(6) PHASE (DEG) - phase in degrees

If a Bode plot has been requested (NBODE = 0), the problem identification information is repeat at the top of a new page and the magnitude and phase plots are given. The magnitude plot is placed on a log-log scale while the phase plot is on a log-linear scale. Phase angles are given in degrees. The plot routine automatically scales the magnitude plot such that the output grid of the magnitude plot is as square as possible. The ordinate scaling of the phase plot is fixed, and the abscissa scaling adjusted to match that of the magnitude plot.

If the polar - Nyquist plot is requested (NYQST = 0), the problem identification is again repeated and the polar plot is displayed.

3.2-4 *EXAMPLE* We wish to determine the frequency response for the transfer function

$$G(s) = \frac{8(0.5 + s)}{4 + 6s + 3s^2 + s^3}$$

for radian frequency range of ω = 0.1 rad/sec to 100 rad/sec. A Bode plot of the magnitude and phase is desired but a Nyquist plot is not needed. We use 100 values of ω logarithmically spaced from 0.1 to 100. The input deck for this example is shown in Table 3.2-3. The numerator polynomial was entered in F format while the denominator was entered in P format.

A portion of the program output is given in Table 3.2-4. Only part of the magnitude plot is shown for lack of space.

Table 3.2-3

Input Data Deck for FRESP Example

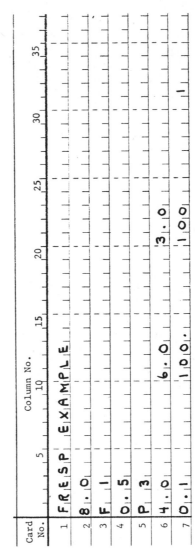

3.2-5 *SUBPROGRAMS USED*   The following subprograms are used by this program:

(1) MAXI
(2) PHNOM
(3) PROOT
(4) PVAL
(5) SEMBL
(6) GRAPH
(7) SPLIT

See Appendix A and Secs. 4.2 and 4.3 for a description of each of the subprograms.

108

Table 3.2-4

Partial Program Output for FRESP Example

FREQUENCY RESPONSE
PROBLEM IDENTIFICATION - FRESP EXAMPLE
****

GAIN = 8.000000E 00

NUMERATOR COEFFICIENTS - IN ASCENDING POWERS OF S

5.000000E-01 1.000000E 00

NUMERATOR ROOTS ARE
REAL PART   IMAG. PART
-5.000000E-01 0.0

DENOMINATOR COEFFICIENTS - IN ASCENDING POWERS OF S

4.000000E 00 6.000000E 00 3.000000E 00 1.000000E 00

DENOMINATOR ROOTS ARE
REAL PART   IMAG. PART
-1.000000E 00 -1.732050E 00
-1.000000E 00 1.732050E 00
-1.000000E 00 0.0

****

RADIAN FREQ.	REAL PART	IMAGINARY PART	MAGNITUDE	PHASE (RAD)	PHASE (DEG)
9.999996E-02	1.014857E 00	4.838795E-02	1.016010E 00	4.764348E-02	2.729771E 00
1.072267E-01	1.017058E 00	5.162836E-02	1.018368E 00	5.071891E-02	2.905980E 00
1.149756E-01	1.019581E 00	5.504400E-02	1.021066E 00	5.393453E-02	3.090220E 00
8.111394E 01	-1.215625E-03	-3.748998E-05	1.216204E-03	-3.110762E 00	-1.782335E 02
8.697583E 01	-1.057320E-03	-3.040770E-05	1.057758E-03	-3.112841E 00	-1.783527E 02
9.326135E 01	-9.196261E-04	-2.466349E-05	9.199572E-04	-3.114779E 00	-1.784637E 02
1.000011E 02	-7.998617E-04	-2.000450E-05	8.001118E-04	-3.116587E 00	-1.785673E 02

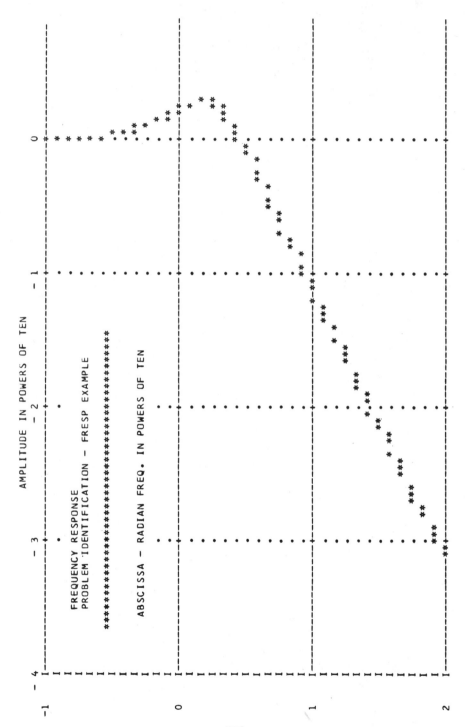

FREQUENCY RESPONSE
PROBLEM IDENTIFICATION - FRESP EXAMPLE

ABSCISSA - RADIAN FREQ. IN POWERS OF TEN

AMPLITUDE IN POWERS OF TEN

```
C   FREQUENCY RESPONSE PROGRAM (FRESP)
C     SUBPROGRAMS USED: MAXI, PHNOM, PROOT, PVAL, SEMBL,
C     GRAPH, SPLIT.
      DATA IPP/1HP/
      DIMENSION A(20),B(20),NAME(5),QPHSR(500),QPHSD(500),
     *  WFREQ(500),QMAG(500),QR(500),QI(500),FREQ(500),
     *  AMAG(500),CO(20),RTR(20),RTI(20)
    1 FORMAT(2X,7(1PE16.6))
   10 FORMAT (5A4)
   11    FORMAT (8E10.0)
   12 FORMAT(2E10.0,4I3)
   13 FORMAT(1H1)
   82 FORMAT(     6X,25HPROBLEM IDENTIFICATION - ,5A4)
  100 FORMAT(1H0,5X,12HRADIAN FREQ.,6X,9HREAL PART,5X,14HIMAGINAE
     *   4X,9HMAGNITUDE,6X,11HPHASE (RAD),5X,11HPHASE (DEG)/)
  900 FORMAT(1H0,5X,22HNUMERATOR COEFFICIENTS
     *   ,27H - IN ASCENDING POWERS OF S      /)
  901 FORMAT (5X,1P6E13.6)
  902 FORMAT(1H0,5X,24HDEMONINATOR COEFFICIENTS
     *   ,27H - IN ASCENDING POWERS OF S      /)
  903 FORMAT (1H1,5X,18HFREQUENCY RESPONSE  )
  904 FORMAT (/3X,45(1H*))
  905 FORMAT(A1,I2)
  906 FORMAT (/6X,19HNUMERATOR ROOTS ARE/8X,9HREAL PART,
     *  8X,10HIMAG. PART)
  907 FORMAT (/6X,21HDENOMINATOR ROOTS ARE/8X,9HREAL PART,
     *  8X,10HIMAG. PART)
  908 FORMAT (/6X,7HGAIN = ,1PE13.6/)
      PI=3.1415926
      PIO2=1.5707963
 1450 READ (5,10,END=1480) (NAME(I),I=1,5)
      PRINT 903
      PRINT 82,NAME
      PRINT 904
      READ 11, GAIN
      PRINT 908, GAIN
      KOUNT=0
   60 KOUNT=KOUNT+1
      READ 905, KEY, N
      NP=N+1
      IF(N.EQ.0) GO TO 65
      IF (KEY.EQ.IPP) GO TO 65
      I=0
   61 I=I+1
      READ 11,RR,RI
      RTR(I)=-RR
      RTI(I)=RI
      IF(RI) 62,63,62
   62 I=I+1
      RTR(I)=-RR
      RTI(I)=-RI
   63 IF(I.LT.N) GO TO 61
      CALL SEMBL(N,RTR,RTI,CO)
      GO TO 67
   65 READ 11  , (CO(I),I=1,NP)
      CO(NP)=1.0
      IF(N) 67,67,66
   66 CALL PROOT (N,CO,RTR,RTI,100)
```

```
 67 GO TO (70,75), KOUNT
 70 DO 71 I=1,NP
 71 A(I)=CO(I)*GAIN
    M=N
    ME=NP
    PRINT 900
    PRINT 901,(CO(I),I=1,NP)
    IF(N.LE.0) GO TO 60
    PRINT 906
    DO 72 I=1,N
 72 PRINT 901, RTR(I),RTI(I)
    GO TC 60
 75 DO 76 I=1,NP
 76 B(I)=CO(I)
    PRINT 902
    PRINT 901, (CO(I),I=1,NP)
    PRINT 907
    DO 77 I=1,N
 77 PRINT 901, RTR(I),RTI(I)
    NE=NP
    PRINT 904
      PRINT 100
    READ 12,WMIN,WMAX,NOMEG,KNOW,NBODE,NYQST
    KNOW=KNOW+1
    WFREQ(1)=WMIN
    IF(NOMEG-2) 41,41,40
 40 XOMEG=NOMEG
    GO TO (50,53,51), KNOW
 51 DWFQ = (WMAX-WMIN)/(XOMEG-1.)
    DO 52 I=2,NOMEG
 52 WFREQ(I)=WFREQ(I-1)+DWFQ
    GO TO 41
 53 READ 11, (WFREQ(I),I=1,NOMEG)
    GO TO 41
 50 Y = (ALOG10(WMAX)-ALOG10(WMIN))/(XOMEG-1.0)
    Z=10.**Y
    DO 42 I=2,NOMEG
    J=I-1
 42 WFREQ(I)=WFREQ(J)*Z
 41 DO 420 K=1,NOMEG
    CALL PVAL (A,M,0.0,WFREQ(K),QNUMR,QNUMI)
    CALL PVAL (B,N,0.0,WFREQ(K),QDENR,QDENI)
      IF (SQRT (QDENI**2 + QDENR**2)) 550, 1550, 550
1550  QMAG(K) = 1.E+20
1551  QR(K) = 1.E+20
1552  QI(K) = 1.E+20
      GO TO 600
 550 QMAG(K)= SQRT(QNUMI**2+QNUMR**2)/SQRT(QDENI**2+QDENR**2)
 551 QR(K)=(QNUMR*QDENR+QNUMI*QDENI)/(QDENR**2+QDENI**2)
 552 QI(K)=(QNUMI*QDENR-QNUMR*QDENI)/(QDENR**2+QDENI**2)
 600  IF (QI(K)) 601, 602, 603
 601  IF (QR(K)) 620, 621, 622
 602  IF (QR(K)) 623, 751, 625
 603  IF (QR(K)) 626, 627, 628
 620  QPHSR(K) =-PI              + ATAN (QI(K)/QR(K))
      GO TO 555
 621  QPHSR(K) = -PIO2
      GO TO 555
 622  QPHSR(K) = -ATAN (-QI(K)/QR(K))
      GO TO 555
 623  QPHSR(K) = PI
      GO TO 555
```

```
 625    QPHSR(K) = 0.0
        GO TO 555
 626    QPHSR(K) = PI              -ATAN (-QI(K)/QR(K))
        GO TO 555
 627    QPHSR(K) = PIO2
        GO TO 555
 628    QPHSR(K) = ATAN (QI(K)/QR(K))
 555    QPHSD(K) = QPHSR(K)*57.2958
        GO TO 420
 751    QPHSR(K) = 1.E+20
        QPHSD(K) = 1.E+20
 420 PRINT 1,WFREQ(K),QR(K),QI(K),QMAG(K),QPHSR(K),QPHSD(K)
     GO TO (1604,1450,1602), KNOW
1604 IF(NBODE) 1601,1601,1602
1601 DO 1600 MA=1,NOMEG
     FREQ(MA)=ALOG10(WFREQ(MA))
     AMAG(MA)=ALOG10(QMAG(MA))
1600 CONTINUE
     MMIN=0
     MMAX=0
     PRINT 903
     PRINT 82,NAME
     PRINT 904
     CALL GRAPH(QPHSD,AMAG,FREQ,NOMEG,MMAX,MMIN)
1602 IF(NYQST) 1603,1603,1450
1603 PRINT 903
     PRINT 82,NAME
     PRINT 904
     CALL SPLIT(QR,QI,NOMEG)
        GO TO 1450
1480    STOP
        END
```

3.3 ROOT LOCUS (RTLOC)

The RTLOC program is used to obtain and plot the zeros of the equation

$$1 + KG(s) = 0$$

as a function of K. Here $G(s)$ is assumed to be a rational function of the form

$$G(s) = K \frac{N(s)}{D(s)}$$

so that we may also consider the problem of obtaining the root locus of the polynomial

$$D_k(s) = D(s) + KN(s)$$

as a function of K.

3.3-1 *THEORY*   If the polynomials $N(s)$ and $D(s)$ are written as

$$N(s) = n_1 + n_2 s + \ldots + n_m s^{m-1} + s^m$$

$$D(s) = d_1 + d_2 s + \ldots + d_n s^{n-1} + s^n$$

where $m < n < 20$, then $D_k(s)$ is given by

$$D_k(s) = Kn_1 + d_1) + (Kn_2 + d_2)s + \ldots + (K + d_{m+1})s^m + \ldots + s^n$$

Now as K vary $D_k(s)$ is computed and the subroutine PROOT is used to factor $D_k(s)$.

An algebraic plus linear progression is used to vary K in the following manner

$$K_{new} = 1.15(K_{old} + 0.05)$$

This procedure has been found to lead to a reasonably uniform spacing of the roots. The use of this procedure assumes that K takes only positive values. If K is to range through negative values, the algebraically smaller (less negative) value must be used as the minimum value. Hence the routine starts at the maximum (less negative) value and becomes increasingly negative until the lower limit is reached. If both positive and negative values are desired, then two separate runs must be made with only positive and only negative values.

When the entire desired range of K has been exhausted, the subroutine SPLIT is used to plot the resulting root locus. This program also has an option which allows one to plot only the portion of the root locus which lies in a specified rectangular region in the s-plane. If this option is selected a rectangular region of the s-plane is specified by giving σ_{max}, σ_{min}, ω_{max} and ω_{min}. Only the portion of the root locus which lies in this region will be plotted. When this option is selected, the progression of gain values is calculated by the rule

$$K_{new} = 1.04(K_{old} + 0.02)$$

This expression leads to a larger number of gain values and hence a more refined plot. The basic use of the option is to refine a portion of the root locus which is of particular interest.

3.3-2 *INPUT FORMAT*   The input deck for the RTLOC program can be divided into three sections: (1) Basic input, (2) Gain range and (3) Option card. The complete input format is summarized in Table 3.3-1 for reference; let us consider each of these three sections separately.

Basic Input: The basic input contains the problem identification and the description of the polynomials N(s) and D(s). Both of the polynomials N(s) and D(s) may be entered in the form of polynomial factors (F mode) or polynomial coefficients (P mode). Note that if the coefficient form is used, the coefficient of the highest order term is assumed to be unity. For simplicity of representation N(s) is shown in P form while D(s) is in F form in Table 3.3-1.

Gain Range: The card after the basic input data gives the range of gain to be investigated. Nominally the range of gain is to include only positive values. If negative gains are desired, the minimum and maximum values should be interchanged on the card. Hence the more negative value will appear in column 11-20. If both positive and negative values are to be used, then two separate runs must be made.

Table 3.3-1

Input Format for RTLOC Program

Card Number	Column Number	Description	Format
1	1-20	Problem identification	5A4
2	1	KEY = P(coefficients supplied)[1]	A1,I2
	2-3	M = order of N(s)	
3	1-10	n_1 = coefficients of N(s)	8E10.0
	11-20	n_2	
	etc.	...	
4	1	KEY = F(polynomial factors supplied)[1]	A1,I2
	2-3	N = order of D(s)	
5	1-10	real part of the second root of D(s)	8E10.0
	11-20	imaginary part of the first root of D(s)	
6	1-10	real part of the second root of D(s)	8E10.0
	11-20	imaginary part of the second root of D(s)	
etc.		...	
Next to last	1-10	KMIN = minimum value of K	8E10.0
	11-20	KMAX = maximum value of K	
Last	1	NOPT = 0 no option	I1, 9X, 4E10.0
		$\neq$ 0 option to be used	
	2-10	NOT USED	
	11-20	σ_{min}	
	21-30	σ_{max}	
	31-40	ω_{min}	
	41-50	ω_{max}	

[1] Both N(s) and D(s) may be supplied in either P or F form.

115

Option Card: The last card in the data deck is used to indicate whether the option which permits one to examine a portion of the root locus which lies in a specified rectangular region. If the option is desired a non-zero integer is placed in column one of the option card and the σ_{min}, σ_{max}, ω_{min} and ω_{max} are entered. If the option is not desired a blank card is used for the last card.

3.3-3 *OUTPUT FORMAT*   Following the problem identification, the numerator and denominator polynomials are described. Both N(s) and D(s) are given in both factored and unfactored form independent of the form that was used to enter the polynomial. The maximum and minimum gain values are given and then the gain values and their associated roots are given in a list. Finally, the subroutine SPLIT is used to form a root locus diagram of the root locations as a function of gain. If a specific region of the s plane has been selected for study, then only the portion of the root locus which lies in this region will be pointed.

3.3-4 *EXAMPLE*   Let us form the root locus for the open-loop transfer function G(s) given by

$$G(s) = \frac{1.2 + s}{s(8 + 9s + s^2)}$$

for K = 0 to K = 30. The input deck for this problem is shown in Table 3.3-2. Here both N(s) and D(s) have been entered is F form.

Table 3.3-2

Input Data Deck for RTLOC Example

Card No.	Column No. 5	10	15	20	25	30
1	RTLOC EXAMPLE ONE					
2	F 1					
3	1.2					
4	F 3					
5	0.0					
6	1.0					
7	8.0					
8	0.0		30.0			
9	(BLANK CARD)					

A portion of the program output for this example is shown in Table 3.3-3. One might wish to rerun the program and investigate more carefully, the root locus around the breakaway and re-entry points.

Table 3.3-3

Partial Program Output for RTLOC Example

```
ROOT LOCUS PROGRAM
PROBLEM IDENTIFICATION - RTLOC EXAMPLE ONE

************************************************

 NUMERATOR COEFFICIENTS IN ASCENDING POWERS OF S

      1.200      1.000

 OPEN-LOOP ZEROES
   REAL PART  IMAG. PART
   -1.200E 00  0.0

 DENOMINATOR COEFFICIENTS IN ASCENDING POWERS OF S

      0.0        8.000      9.000      1.000

 OPEN-LOOP POLES
   REAL PART  IMAG. PART
   -0.0         0.0
   -1.000E 00   0.0
   -8.000E 00   0.0

 MIN. GAIN =   0.0                    MAX. GAIN =   3.00E 01

************************************************

     1      GAIN =  0.0

        ROOTS ARE
   REAL PART  IMAG. PART

   -1.000E 00   0.0
   -8.000E 00   0.0
    0.0         0.0

     2      GAIN =  5.750E-02
```

```
    31      GAIN =  2.500E 01

        ROOTS ARE
   REAL PART  IMAG. PART

   -3.846E 00 -2.854E 00
   -3.846E 00  2.854E 00
   -1.308E 00   0.0

    32      GAIN =  2.880E 01

        ROOTS ARE
   REAL PART  IMAG. PART

   -3.857E 00 -3.465E 00
   -3.857E 00  3.465E 00
   -1.286E 00   0.0
```

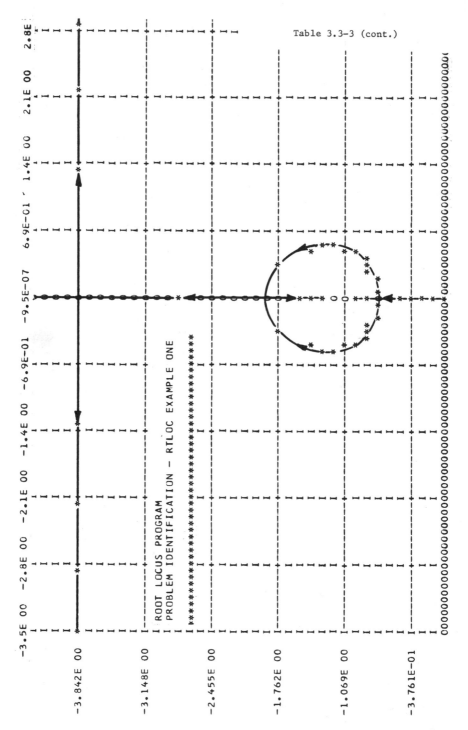

Table 3.3-3 (cont.)

ROOT LOCUS PROGRAM
PROBLEM IDENTIFICATION — RTLOC EXAMPLE ONE

118

3.3-5 *SUBPROGRAMS USED*   The following subprograms are used by this
program

     (1)   PROOT
     (2)   SEMBL
     (3)   SPLIT

See Appendix A and Sec. 4.3 for a description of each of these subprograms.

3.3-6 *PROGRAM LISTING*

```
C   ROOT LOCUS PROGRAM (RTLOC)
        DIMENSION GHN(11),GHD(11),A(11),U1(999),V1(999),
     +  U(10),V(10),REAL(2),EMAJ(2),U2(999),V2(999),
     +  NAME(5)
        DATA IPP/1HP/
  1 FORMAT (5A4)
  2 FORMAT(8E10.0)
  6 FORMAT(1H1,4X,18HROOT LOCUS PROGRAM  / 5X,
     *   25HPROBLEM IDENTIFICATION - ,5A4//3X,45(1H*))
  7 FORMAT(1I1,9X,4E10.0)
  8 FORMAT(/,23H OPTION HAS BEEN TAKEN )
 11 FORMAT(/12H SIGMA MIN =,1PE11.2,4X,12H SIGMA MAX =,
     +  1PE11.2/12H OMEGA MIN =,1PE11.2,4X,12H OMEGA MAX =,
     +  1PE11.2)
 22 FORMAT (/5X,I3,5X,7H GAIN = ,1PE11.3//12X,9HROOTS ARE/
     +  7X,9HREAL PART,2X,10HIMAG. PART/)
 23 FORMAT(5X,10(1PE11.3))
225 FORMAT(/5X,10F10.3)
230 FORMAT( / 5X,23HNUMERATOR COEFFICIENTS    ,
     *   24HIN ASCENDING POWERS OF S    )
231 FORMAT( / 5X,25HDENOMINATOR COEFFICIENTS   ,
     *   24HIN ASCENDING POWERS OF S   )
300 FORMAT(1H0,2X,45(1H*))
306 FORMAT (/5X,12HMIN. GAIN = ,1PE10.2,10X,13H MAX. GAIN = ,
     *   1PE10.2)
505 FORMAT (/5X,16HOPEN-LOOP ZEROES/7X,9HREAL PART,
     *  2X,10HIMAG. PART)
506 FORMAT (/5X,15HOPEN-LOOP POLES/7X,9HREAL PART,
     *  2X,10HIMAG. PART)
800 FORMAT (A1,I2)
250 CONTINUE
        READ (5,1,END=900) (NAME(I),I=1,5)
        PRINT 6,(NAME(I),I=1,5)
        DO 400 I=1,10
        GHN(I)=0.0
        GHD(I)=0.0
        A(I)=0.0
        U(I)=0.0
400 V(I)=0.0
        DO 401 I=1,999
        V1(I)=0.0
        U1(I)=0.0
        V2(I)=0.0
401 U2(I)=0.0
        KOUNT=0
500 KOUNT=KOUNT+1
        READ 800,KEY,N
        NP=N+1
        IF(KEY.EQ.IPP) GO TO 804
        IF(N.EQ.0) GO TO 804
        I=0
801 I=I+1
        READ 2, RREAL,RIMAG
```

119

```
      U(I)=-RREAL
      V(I)=RIMAG
      IF(RIMAG) 802,803,802
802   I=I+1
      U(I)=-RREAL
      V(I)=-RIMAG
803   IF(I.LT.N) GO TO 801
      CALL SEMBL(N,U,V,GHD)
      GO TO (807,810), KOUNT
804   READ 2, (GHD(I),I=1,NP)
      GHD(NP)=1.0
      IF(N) 806,806,805
805   CALL PROOT(N,GHD,U,V,+100)
806   GO TO(807,810), KOUNT
807   DO 808 I=1,NP
808   GHN(I)=GHD(I)
      NN=NP
      PRINT 230
      PRINT 225,(GHN(I),I=1,NP)
      IF(N.LE.0) GO TO 500
      PRINT 505
      DO 809 I=1,N
809   PRINT 23,U(I),V(I)
      GO TO 500
810   ND=NP
      PRINT 231
      PRINT 225,(GHD(I),I=1,ND)
      PRINT 506
      DO 811 I=1,N
811   PRINT 23,U(I),V(I)
      READ 2, GMIN, GMAX
      PRINT 306, GMIN, GMAX
      IF(GMAX)3000,900,301
3000  SIGNG=-1.
      GO TO 302
301   SIGNG=+1.
302   CONTINUE
      READ 7,NOPT,(REAL(I),I=1,2),(EMAJ(I),I=1,2
      IF(NOPT)9,9,10
10    PRINT 300
      PRINT 8
      PRINT 11,REAL(1),REAL(2),EMAJ(1),EMAJ(2)
9     PRINT 300
      IR=1
      ITER=0
      G=GMIN
      N=ND-1
24    DO 20 I=1,ND
20    A(I)=GHN(I)*G+GHD(I)
      CALL PROOT(N,A,U,V,IR)
      DO 21 I=1,N
      J=I+ITER*N
      U1(J)=U(I)
21    V1(J)=V(I)
      ITER=ITER+1
      PRINT 22,ITER,G
      DO 25 I=1,N
25    PRINT 23,U(I),V(I)
      IF(NOPT)227,227,228
227   G=1.15*(G+SIGNG*0.05)
      GO TO 229
228   G=1.04*(G+SIGNG*0.02)
229   IF(G-GMAX)26,26,27
```

120

```
   26  GO TO 24
   27  IF (NOPT)28,28,29
   29  NPTS=0
       MT=ITER*N
       DO 31 I=1,MT
       IF(U1(I)-REAL(1))31,30,30
   30  IF(REAL(2)-U1(I))31,32,32
   32  IF(V1(I)-EMAJ(1))31,33,33
   33  IF(EMAJ(2)-V1(I))31,34,34
   34  NPTS=NPTS+1
       U2(NPTS)=U1(I)
       V2(NPTS)=V1(I)
   31  CONTINUE
       PRINT 6, (NAME(I),I=1,5)
       PRINT 230
       PRINT 225,(GHN(I),I=1,NN)
       PRINT 231
       PRINT 225,(GHD(I),I=1,ND)
       PRINT 306, GMIN, GMAX
       PRINT 300
       CALL SPLIT(U2,V2,NPTS)
       GO TO 35
   28  NPTS=ITER*N
       PRINT 6, (NAME(I),I=1,5)
       PRINT 230
       PRINT 225,(GHN(I),I=1,NN)
       PRINT 231
       PRINT 225,(GHD(I),I=1,ND)
       PRINT 306, GMIN, GMAX
       PRINT 300
       CALL SPLIT(U1,V1,NPTS)
   35  CONTINUE
       GO TO 250
  900  STOP
       END
```

3.4 PARTIAL FRACTION EXPANSION (PRFEXP)

The PRFEXP program is used to determine the partial fraction expansion of ratio of two polynomials. The denominator polynomial may have multiple real roots but complex roots are assumed to be simple.

3.4-1 *THEORY* The PRFEXP program is based on an algorithm given by Pottle[1]. Let the rational function of interest be represented by

$$G(s) = K \frac{A(s)}{B(s)}$$

where

$$A(s) = a_1 + a_2 s + \ldots + a_m s^{m-1} + s^m$$

$$B(s) = b_1 + b_2 s + \ldots + b_n s^{n-1} + s^n$$

with $m < n \leq 9$. We represent the denominator polynomial in factoral form as

$$B(s) = \prod_{i=1}^{L} (s - s_i)^{m_i}$$

where L is the number of distinct root of $B(s)$ and m_i is the multiplicity of the ith root of $B(s)$. The rational function $G(s)$ is to be represented in partial fraction form as

$$G(s) = \sum_{i=1}^{L} \sum_{j=1}^{m_i} \frac{R_{ij}}{(s - s_i)^j}$$

In other words R_{ij} is the numerator coefficient of the jth-power-term associated with the ith root in the partial fraction expansion. The PRFEXP program determines the values of these numerator coefficients.

We begin by extracting the highest power term of each root so that $G(s)$ becomes

$$G(s) = \sum_{i=1}^{L} \frac{R_{im_i}}{(s - s_i)^{m_i}} + \frac{P(s)}{Q(s)}$$

where R_{im_i} may be evaluated from

$$R_{im_i} = \frac{KA(s)}{\displaystyle\prod_{\substack{j=1 \\ j \neq i}}^{L} (s - s_j)} \Bigg|_{s = s_i}$$

and

$$P(s) = KA(s) - \sum_{i=1}^{L} \prod_{\substack{j=1 \\ j \neq i}}^{L} R_{im_i} (s - s_i)^{m_i}$$

---

[1] Pottle, C., "On the Partial Fraction Expansion of a Rational Function with Multiple Poles by Digital Computer," *IEEE Trans. Circuit Theory*, Vol. CT-11, pp. 161-162, March 1964.

$$Q(s) = \prod_{i=1}^{L} (s - s_i)^{m_i - 1}$$

If the common factor $\prod_{i=1}^{L} (s - s_i)$ is divided out of $P(s)$ and $Q(s)$ a new rational function $G_1(s)$ is generated so that

$$G(s) = \sum_{i=1}^{L} \frac{R_{im_i}}{(s - s_i)^{m_i}} + G_1(s)$$

The function $G_1(s)$ has a denominator polynomial of order n-L and all of the roots have had their multiplicity reduced by one. If a root was simple in $G(s)$, it does not appear in $G_1(s)$. This algorithm is repeated until all multiplicities have been reduced to zero. The results take the form of an L by max$\{m_i\}$ matrix of numerator coefficients. For a simple root only the element R_{11} will be non-zero.

Roots are assumed to be identical if their real and imaginary parts do not differ by more than 0.005. There can be no more than nine distinct roots and the maximum multiplicity is also nine.

3.4-2 *INPUT FORMAT* The input for the PRFEXP program consists of problem identification and the description of the rational function $G(s)$. This input is identical to the basic input of the FRESP program. Both the numerator and denominator polynomials may be entered as either polynomial factors (KEY = F) or as polynomial coefficients (KEY = P). Note that the coefficient of the highest order term is assumed to be unity. The input format for the PRFEXP program is summarized in Table 3.4-1 for easy reference. In that table, the polynomial A(s) is shown in the P mode while B(s) is shown in the F mode. Note that multiple roots must be entered once for each time that they are to appear.

Table 3.4-1

Input Format for the PRFEXP Program

Card Number	Column Number	Description	Format
1	1-20	Problem identification	5A4
2	1-10	K = gain	8F10.3
3	1	KEY = P (coefficients supplied)[1]	A1,I2
	2-3	M = order of A(s) polynomial	
4	1-10	a_1 = coefficients of A(s)	8F10.3
	11-20	a_2	
	etc.	...	
5	1	KEY = F (polynomial factors supplied)[1]	A1,I2
	2-3	N = order of B(s) polynomial	
6	1-10	real part of the first root of B(s)	8F10.0
	11-20	imaginary part of the first root of B(s)	
7	1-10	real part of the second root of B(s)	8F10.0
	11-20	imaginary part of the second root of B(s)	
etc.		...	

[1]Both A(s) and B(s) may be supplied in either P or F form.

123

3.4-3 *OUTPUT FORMAT*   The problem identification and polynomial description are repeated for reference. Both the numerator and denominator polynomials
are given in factored as well as unfactored form independent of which mode was
used to enter them. In the case of the denominator roots, multiple roots are
listed only once; the multiplicity of each root is also given.

The numerator coefficient array for the partial fraction expansion is
given next. The numerator coefficients are listed in the same order as the
denominator roots. The first coefficient in any row is the coefficient of the
first-order term; the second coefficient is for the second-order term and so
forth.

3.4-4 *EXAMPLE*   Let us consider the determination of the partial fraction
expansion of

$$G(s) = \frac{12(3 + s)}{s^2(s + 2)^2[(s + 3)^2 + 3^2]}$$

In this problem, we have two denominator roots which are second-order and a
pair of complex conjugate roots. The input data deck for this example is shown
in Table 3.4-2.

Table 3.4-2

Input Data Deck for PRFEXP Example

Card No.	Column No. 5	10	15	20	25	30
1	PRFEXP EXAMPLE					
2	12.0					
3	P					
4	3.0					
5	F 6					
6	0.0					
7	0.0					
8	2.0					
9	2.0					
10	3.0	3.0				

The program output for this example is given in Table 3.4-3. From this
output we see that the partial fraction expansion of G(s) is

$$G(s) = \frac{-0.02 + j0.027}{s + 3 + j3} + \frac{-0.02 - j0.027}{s + 3 - j3} + \frac{0.54}{s + 2} + \frac{0.3}{(s + 2)^2} + \frac{-0.5}{s} + \frac{0.5}{s^2}$$

124

Table 3.4-3

Program Output for PRFEXP Example

PARTIAL FRACTION EXPANSION
PROBLEM IDENTIFICATION -   PRFEXP EXAMPLE

NUMERATOR GAIN = 1.200E 01

NUMERATOR COEFF. - IN ASCENDING POWERS

 3.000E 00 1.000E 00

NUMERATOR ROOTS

  REAL PART   IMAG. PART
-3.0000000E 00 0.0

DENOMINATOR COEFF. - IN ASCENDING POWERS

 0.0   0.0   7.200E 01 9.600E 01 4.600E 01 1.000E 01 1.00

****

DENOMINATOR ROOTS

  REAL PART   IMAG. PART MULTIPLICITY
-3.0000000E 00 3.0000000E 00   1
-3.0000000E 00-3.0000000E 00   1
-2.0000000E 00 0.0   2
-0.0   0.0   2

****

RESIDUE MATRIX - REAL PART

-2.0000000E-02
-2.0000000E-02
 5.3999996E-01 2.9999995E-01
-4.9999976E-01 5.0000000E-01

RESIDUE MATRIX - IMAG. PART

-2.6666664E-02
 2.6666664E-02
-0.0   -0.0
-0.0   -0.0

****

   3.4-5 *SUBPROGRAMS USED* The following subprograms are used by this program

(1)	DIVP	(5)	PFEXP	(9)	SEMBL
(2)	NORMP	(6)	PMUL	(10)	SORT
(3)	PADD	(7)	PROOT	(11)	SUBP
(4)	PEXCG	(8)	PVAL		

See Appendix A for a description of each of these subprograms.

```
C    PARTIAL FRACTION EXPANSION PROGRAM (PRFEXP)
C    SUBPROGRAMS USED:   DIVP, NORMP, PADD, PEXCG, PFEXP,
C       PMUL, PROOT, PVAL, SEMBL, SORT, SUBP.
         DIMENSION A(10), B(10), M(10), RR(10), RI(10), CR(10)
         DIMENSION RESR(10,10), RESI(10,10) ,NAME(5),CI(10)
         DATA IPP/1HP/
         EPSLN=0.001
      2 FORMAT (8F10.3)
     14 FORMAT (40X, 28H YOU'VE GOT TO BE KIDDING---  ,/,
        +    40X, 35H YOU HAVE A MULTIPLE COMPLEX POLE-   ,/,
        +    40X, 8H I QUIT            )
     17 FORMAT (10(8X,I2))
   2000 FORMAT (/ ,5X,27HRESIDUE MATRIX - REAL PART    /)
   2001 FORMAT (5X,6(1PE14.7))
   2002 FORMAT ( / ,5X,28HRESIDUE MATRIX - IMAG. PART    / )
   2003 FORMAT (/ ,5X,21HNUMERATOR COEFF. - IN ,
        +   17H ASCENDING POWERS  ,/)
   2004 FORMAT (/ 5X,23HDENOMINATOR COEFF. - IN ,
        +   17H ASCENDING POWERS  ,/)
   2005 FORMAT (/5X,17HNUMERATOR GAIN = ,1PE10.3)
   2006 FORMAT ( / ,5X,17HDENOMINATOR ROOTS //,7X,9HREAL PART,
        +   5X,10HIMAG. PART,2X,12HMULTIPLICITY     )
   2007 FORMAT (A1,I2)
   2008 FORMAT (5X,1PE14.7,1PE14.7,I6)
   2009 FORMAT (5A4)
   2010 FORMAT (      5X,25HPROBLEM IDENTIFICATION -   ,2X,5A4)
   2011 FORMAT (1H1,4X,26HPARTIAL FRACTION EXPANSION   )
   2012 FORMAT (/5X,45(1H*))
   2013 FORMAT (5X,1P8E10.3)
   2014 FORMAT (/5X,15HNUMERATOR ROOTS//7X,9HREAL PART,
        *   8X,10HIMAG. PART)
   2015 FORMAT (5X,1PE14.7,1PE14.7)
    103 READ (5,2009,END=102) (NAME(I),I=1,5)
         PRINT 2011
         PRINT 2010,(NAME(I),I=1,5)
         DO 100 I=1,10
         DO 99 J = 1, 10
         RESI(I,J) = 0.
         RESR(I,J) = 0.
     99 CONTINUE
         A(I) = 0.
         B(I) = 0.
         M(I) = 0
         RR(I) = 0.
         RI(I) = 0.
         CR(I) = 0.
         CI(I) = 0.
    100 CONTINUE
         READ 2, GAIN
         PRINT 2005, GAIN
         KOUNT=0
    400 KOUNT=KOUNT+1
         READ 2007, KEY, N
         NP=N+1
         IF(KEY.EQ.IPP) GO TO 404
         IF(N.EQ.0) GO TO 404
         I=0
    401 I=I+1
```

126

```
      READ 2, RREAL,RIMAG
      RR(I)=-RREAL
      RI(I)=RIMAG
      IF(RIMAG) 402,403,402
  402 I=I+1
      RR(I)=-RREAL
      RI(I)=-RIMAG
  403 IF(I.LT.N) GO TO 401
      CALL SEMBL (N,RR,RI,B)
      GO TO (407,410), KOUNT
  404 READ 2,(B(I),I=1,NP)
      B(NP)=1.0
      IF(N) 406,406,405
  405 CALL PROOT(N,B,RR,RI,+100)
  406 GO TO (407,410), KOUNT
  407 DO 408 I=1,NP
  408 A(I)=B(I)
      NNUM=NP
      NNM=N
      PRINT 2003
      PRINT 2013,(A(I),I=1,NNUM)
      IF(N.LE.0) GO TO 400
      PRINT 2014
      DO 409 I=1,N
  409 PRINT 2015,RR(I),RI(I)
      GO TO 400
  410 NDEN=NP
      NDM=N
      PRINT 2004
      PRINT 2013,(B(I),I=1,NP)
      DO 411 I=1,NNUM
  411 A(I)=A(I)*GAIN
      IT = 0
      DO 3 I=1,NDM
      IF (IT) 6,6,20
   20 DO 4 J = 1, IT
      IF (ABS(RR(I)-CR(J))-EPSLN) 5,5,4
    5 IF(ABS(RI(I)-CI(J))-EPSLN) 3,3,4
    4 CONTINUE
    6 IT = IT +1
      CR(IT) = RR(I)
      CI(IT) = RI(I)
      DO 7 J=1,NDM
      IF(ABS(CR(IT)-RR(J))-EPSLN) 8,8,7
    8 IF(ABS(CI(IT)-RI(J))-EPSLN) 9,9,7
    9 M(IT) = M(IT) +1
    7 CONTINUE
    3 CONTINUE
C          SORT ROOTS
      CALL SORT (CR, CI, IT,M)
      PRINT 2012
      PRINT 2006
      PRINT 2008,(CR(I),CI(I),M(I),I=1,IT)
      DO 11 I = 1, IT
      IF (CI(I)) 12, 11, 12
   12 IF(M(I) -2) 11,13,13
   13 PRINT 14
      GO TO 102
   11 CONTINUE
      CALL PFEXP (A,NNM,CR,CI,M,IT,RESR,RESI)
      PRINT 2012
  205 PRINT 2000
      DO 209 I=1,IT
```

```
      IJL=M(I)
209   PRINT 2001, (RESR(I,J),J=1,IJL)
      PRINT 2002
      DO 210 I=1,IT
      ILL=M(I)
210   PRINT 2001,(RESI(I,J),J=1,ILL)
      PRINT 2012
      GO TO 103
102   STOP
      END
```

4 DESIGN PROBLEMS

4.1 INTRODUCTION

In this chapter, we consider three design problems which illustrate some
of the ways in which the programs of the two preceding chapters can be used
to analyze and design linear control and communication systems. Section 4.2
discusses the use of STVARFDBK, LUEN and SERCOM to design a simplified
vertical channel landing condition autopilot. In this problem, the angle
of attack is assumed to be inaccessible. The next section, the program
STVARFDBK is used to design a state variable feedback control for a simple
nuclear reactor model. SENSIT is then used to analyze the resulting system.
Section 4.4 considers a communication theory problem in which the program
RICATI is used to design a Wiener filter.

4.2 DESIGN OF A SIMPLIFIED VERTICAL CHANNEL LANDING CONDITION AUTOPILOT

4.2-1 *INTRODUCTION*   This section illustrates the use of STVARFDBK, LUEN, and SERCOM in the design of a simplified vertical channel landing condition autopilot.

4.2-2 *PROBLEM STATEMENT*   A set of simplified equations describing the airplane vertical channel, after linearization, are [1]

$$\dot{\theta}_A = q$$

$$\dot{q} = K_2[C_{M\alpha}\alpha + C_{M\delta}\delta_e + K_1 C_{Mq}q]$$

$$\dot{\alpha} = q + K_3[C_{L\alpha}\alpha + C_{L\delta}\delta_e]$$

$$\dot{\delta}_e = -20\delta_e + 20u$$

where: q = pitch rate of airplane (rad/sec)

θ_A = pitch angle of airplane (rad)

$K_1 = C/2V_T$

$K_2 = QSC/I_y$

$K_3 = -SQ/mV_T$

Q = dynamic pressure (lb/sq ft)

S = effective wing area (sq ft)

C = mean aerodynamic chord (ft)

m = airplane mass·(lb/ft/sec^2)

I_y = moment of inertia of airplane about lateral axis

α = angle of attack (rad)

δ_e = elevator deflection (rad)

V_T = airplane speed (ft/sec)

u = control input

$C_{M\alpha}$ = change in pitching moment coefficient with varying angle of attack (1/rad)

$C_{M\delta}$ = change in pitching moment coefficient with varying elevator deflection (1/rad)

C_{Mq} = change in pitching moment with varying pitch velocity (1/rad)

$C_{L\alpha}$ = change in lift coefficient with varying angle of attack (1/rad)

$C_{L\delta}$ = change in lift coefficient with changes in elevator deflection

Defining $x^T = (\theta_A, q, \alpha, \delta_e)$, we obtain the following state variable formulation of the problem

$$\dot{x} = \begin{bmatrix} 0 & 1 & 0 & 0 \\ 0 & K_1 K_2 C_{M\delta} & K_2 C_{M\alpha} & K_2 C_{M\delta} \\ 0 & 1 & K_3 C_{L\alpha} & K_3 C_{L\delta} \\ 0 & 0 & 0 & -20 \end{bmatrix} x + \begin{bmatrix} 0 \\ 0 \\ 0 \\ 20 \end{bmatrix} u$$

---

[1] *Dynamics of the Airframe*, Bauer Rept. AE-61-4(II), Vol. II, Sept. 1952.

$$y = [1 \quad 0 \quad 0 \quad 0] \, \underset{\sim}{x}$$

Based on an altitude of 500 feet, approach speed of 260 knots and typical values for the various parameters, the $\underset{\sim}{A}$ matrix can be evaluated as

$$\underset{\sim}{A} = \begin{bmatrix} 0 & 1 & 0 & 0 \\ 0 & -0.856 & -2.737 & -8.212 \\ 0 & 1 & -0.521 & -0.077 \\ 0 & 0 & 0 & -20 \end{bmatrix}$$

It is desired to obtain a feedback control law for this system which yields a closed-loop transfer function having poles at $s = -0.5$, $s = -1.5 \pm j1.5$, $s = -20$. Due to physical constraints, the angle of attack $\alpha = x_3$ is unavailable for measurement and must be estimated or observed from x_1, x_2, and x_4 if x_3 is required by the feedback control law.

4.2-3 *METHOD OF SOLUTION*   Using the input deck shown in Fig. 4.2-1 in STVARFDBK we find that the open-loop transfer function is given by

$$G_p(s) = \frac{-164.24(s+0.495)}{s(s+20)(s+0.69+j1.65)(s+0.69-j1.65)}$$

and that a state-variable feedback control given by

$$\underset{\sim}{k}^T = [1.0 \quad 0.472 \quad -0.658 \quad -0.192]$$

$$K = -0.553$$

will produce a closed-loop transfer function

$$G_c(s) = \frac{90.825(s+0.495)}{(s+0.5)(s+20)(s+1.5+j1.5)(s+1.5-j1.5)}$$

Obviously the indicated feedback gain vector $\underset{\sim}{k}^T$ requires that x_3 be measurable. We will use both LUEN and SERCOM to overcome this difficulty.

In the previously used terminology, the measurable states are given by $\underset{\sim}{y}^* = \underset{\sim}{C}^* \underset{\sim}{x}$, where

$$\underset{\sim}{C}^* = \begin{bmatrix} 1 & 0 & 0 & 0 \\ 0 & 1 & 0 & 0 \\ 0 & 0 & 0 & 1 \end{bmatrix}$$

Using the input deck shown in Fig. 4.2-2 as input to OBSERV, it is verified that the observability index for the system is given by $\nu = 2$; therefore, a first-order observer is sufficient for the system.

First we consider designing a feedback compensator using LUEN. For this problem, the matrix F is actually a scalar which we choose to be $F = -5.0$ which coincides with none of the eigenvalues of $\underset{\sim}{A}$. Using the input deck of Fig. 4.2-3 in LUEN, we find that the compensator is given by

$$\dot{z} = -5z - 5.12y_2^* + 11.86u$$

131

and the control is

$$u = -0.553(r - y_1^* - 1.55y_2^* + 0.785y_3^* - z)$$

Fig. 4.2-1

Input Deck for STVARFDBK

Card No.	Column 5-10	Column 10-15	Column 20-25	Column 30-35
1	AUTOPILOT	DESIGN	4	
2	0.0	1.0	0.0	0.0
3	0.0	-.856	-2.737	-8.212
4	0.0	1.0	-.521	-1.077
5	0.0	0.0	0.0	-20.
6	0.0	0.0	0.0	20.
7	1.0			
8				
9	F			
10	0.5			
11	20.			
12	1.5	1.5		

Fig. 4.2-2

Input Deck for OBSERV

Card No.	Column 5-10	Column 10-15	Column 20-25	Column 30-35
1	AUTOPILOT	DESIGN	4 3	
2	0.0	1.0	0.0	0.0
3	0.0	-0.856	-2.737	-8.212
4	0.0	1.0	-0.521	-0.077
5	0.0	0.0	0.0	-20.0
6	1.0	0.0	0.0	0.0
7	0.0	1.0	0.0	0.0
8	0.0	0.0	0.0	1.0

Now we consider the same problem using SERCOM. Again, the required compensator order is one. As described in the discussion of SERCOM, we choose $D = 0$, $f = 1 = e$, and obtain the augmented system

$$\dot{\underset{\sim}{\hat{x}}} = \left[\begin{array}{c|c} \underset{\sim}{A} & \underset{\sim}{b} \\ \hline 0 & D \end{array}\right] \underset{\sim}{\hat{x}} + \begin{bmatrix} 0 \\ 0 \\ 0 \\ 0 \\ 1 \end{bmatrix} w$$

We choose a desired fifth-order transfer function of

$$\frac{y(s)}{r(s)} = \frac{675(s+0.495)}{(s+0.5)(s+20)(s+15)(s+1.5+j1.5)(s+1.5-j1.5)}$$

and solve for

$$K = -8.297$$

$$\underset{\sim}{k}^T = [1.0 \quad 0.468 \quad -0.721 \quad -0.191 \quad -2.064]$$

using STVARFDBK. The resulting input deck for SERCOM is shown in Fig. 4.2-4, and the following output is obtained.

THE COMPENSATOR SYSTEM MATRIX

1.6897094+01

MINOR LOOP FEEDBACK COEFFICIENTS

4.8449613-01   -2.6342711-01   8.9388296-02

MAJOR LOOP FEEDBACK COEFFICIENTS

7.1865765+00   -5.6291422+00   1.3259050+00

Fig. 4.2-3

Input Deck for LUEN

Card No.	Col 1	Col ~10	Col ~18	Col ~30
1	AUTOPILOT DESIGN		4 3 1	
2	0.0	1.0	0.0	0.0
3	0.0	-0.856	-2.737	-8.212
4	0.0	1.0	-0.521	-0.077
5	0.0	0.0	0.0	-20.0
6	0.0	0.0	0.0	20.0
7	1.0	0.0	0.0	0.0
8	0.0	1.0	0.0	0.0
9	0.0	0.0	0.0	1.0
10	1.0	0.472	-0.658	-0.192
11	F			
12	5.0			

Fig. 4.2-4

Input Deck for SERCOM

Column No.

Card No.	5	10	15	20	25	30	35	40	45
1	AUTOPILOT DESIGN								
2	01.0			0.0	4.3				
3	01.0	-0.856		0.0	-3.737	-8.212			
4	01.0	1.0		-0.521		-10.077			
5	01.0	0.0		0.0		-20.0			
6	01.0	0.0		0.0		20.0			
7	1.0	0.0		0.0		0.0			
8	01.0	1.0		0.0		0.0			
9	01.0	0.0		0.0		1.0			
10	0.0								
11	1.0								
12	1.0								
13	1.0	0.468		-0.721		-0.191		-2.064	
14	-8.297								

4.3 STATE VARIABLE FEEDBACK CONTROL OF NUCLEAR REACTOR SYSTEMS

4.3-1 *INTRODUCTION* This section contains two examples using STVARFDBK, and SENSIT in the design of state variable feedback controls for simple nuclear reactor models.

4.3-2 *PROBLEM FORMULATION; EXAMPLE ONE* A simplified and normalized model of a two-core reactor with no delayed neutrons or temperature feedback is given by

$$\dot{x}_1 = -10x_1 + 10x_2 + x_3$$

$$\dot{x}_2 = -10x_2 + 10x_1$$

$$\dot{x}_3 = -10x_3 + u$$

$$y = x_1 + x_2$$

Here the coupling coefficient is assumed to be 10 sec^{-1}. x_1 is the power in core one, x_2 the power in core two. Only core one is controlled through the controller with time constant 0.1. A block diagram of the system is shown in Fig. 4.3-1.

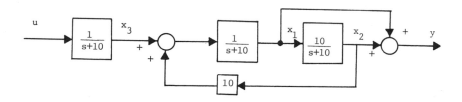

Fig. 4.3-1 Block Diagram of the Two-Core Reactor of Example One.

The desired closed-loop transfer function has poles given by $s = -20$, $s = -10 \pm j10$.

4.3-3 *PROBLEM SOLUTION* We will use STVARFDBK to find the control law $u = K(r - \underline{k}^T x)$ which gives the desired closed-loop poles. The input deck for STVARFDBK is shown in Fig. 4.3-2. The required control law is given by

$$K = 200$$

$$\underline{k}^T = [1.0 \quad 1.0 \quad 0.05]$$

To check the behavior of the closed-loop system, its root locus is plotted versus K by SENSIT. The input deck is given in Fig. 4.3- and the resulting root locus shown in Fig. 4.3-4. It may be observed that the system is stable for all values of K.

135

Fig. 4.3-2

Input Deck for STVARFDBK

Card No.	Columns 1–10	Columns 11–20	Columns 21–25
1	REACTOR CONTROL		3
2	-10.0	10.0	1.0
3	10.0	-10.0	0.0
4	0.0	0.0	-10.0
5	0.0	0.0	1.0
6	1.0	1.0	
7			
8	F		
9	20.0		
10	10.0	10.0	

Fig. 4.3-3

Input Deck for SENSIT

Card No.	Columns 1–10	Columns 11–20	Columns 21–30	Columns 31–35
1	REACTOR CONTROL		3	
2	-10.0	10.0	1.0	
3	10.0	-10.0	0.0	
4	0.0	0.0	-10.0	
5	0.0	0.0	1.0	
6	1.0	1.0	0.0	
7	1.0	1.0	0.05	
8	200.0			
9	0.0	0.0	0.0	
10	0.0	1.0	.0001	100
11	Y			

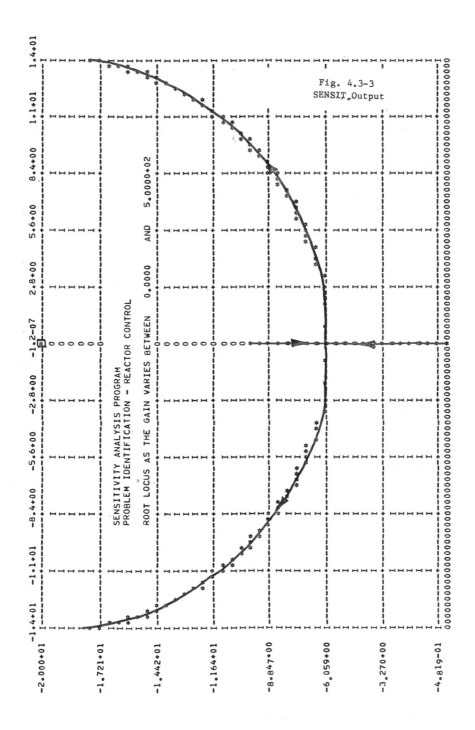

Fig. 4.3-3
SENSIT Output

SENSITIVITY ANALYSIS PROGRAM
PROBLEM IDENTIFICATION — REACTOR CONTROL

ROOT LOCUS AS THE GAIN VARIES BETWEEN   0.0000   AND   5.0000+02

4.3-4 *PROBLEM FORMULATION; EXAMPLE TWO*   The equations for two coupled cores with negative temperature reactivity effects, identical power levels, and with delayed neutrons neglected are

$$\dot{x}_1 = - \frac{\alpha}{\ell} x_1 + \frac{\alpha}{\ell} x_2 - \frac{n_o \gamma}{\ell} x_3 + u$$

$$\dot{x}_2 = - \frac{\alpha}{\ell} x_2 + \frac{\alpha}{\ell} x_1 - \frac{n_o \gamma}{\ell} x_4$$

$$\dot{x}_3 = A x_1 - \omega x_3$$

$$\dot{x}_4 = A x_2 - \omega x_4$$

$$y = x_1 + x_2$$

Here x_1 and x_2 are the changes in power of the controlled and uncontrolled cores respectively.   Rough parameter values are

$$\frac{\alpha}{\ell} = 10 \ \text{sec}^{-1} \qquad\qquad \frac{n_o \gamma}{\ell} = 10 \ \frac{MW}{{}^\circ R \ sec}$$

$$A = 1 \ \frac{{}^\circ R}{MW \ sec} \qquad\qquad \omega = 1 \ \text{sec}^{-1}$$

The desired $y(s)/r(s)$ has closed-loop poles at $s = -1$, $s = -1.54$, $s = -19.5$ and $s = -50$.

4.3-5 *PROBLEM SOLUTION*   The reader should use STVARFDBK to verify the following solution :

$$K = 50$$

$$\underline{k}^T = [1 \quad\quad 1 \quad\quad -0.2 \quad\quad -0.2]$$

4.4   A COMMUNICATION PROBLEM

4.4-1   *INTRODUCTION*   This section discusses a communication theory problem
which can be easily solved using RICATI.

4.4-2   *PROBLEM FORMULATION*   A signal is generated by passing white noise
through a linear system with transfer function $H(s) = (s+4)/s(s+2)(s+3)$ as
in Fig. 4.4-1.

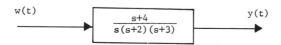

Fig. 4.4-1   Signal Model

The signal $y(t)$ is passed through two different channels each of whose outputs
is corrupted by an additive white noise as in Fig. 4.4-2.   The channel outputs
$z_1(t)$ and $z_2(t)$ are available for measurement.   An estimate of $y(t)$ is desired.

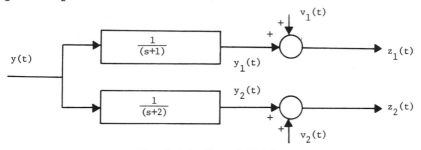

Fig. 4.4-2   Channel Model

The pertinent signal statistics are the following:

$$E\{w(t)\} = E\{v_i(t)\} = 0 \qquad\qquad i = 1,2$$

$$Cov\{w(t)w(\tau)\} = \cdot\delta(t-\tau)$$

$$Cov\{v_1(t)v_1(\tau)\} = \delta(t-\tau)$$

$$Cov\{v_2(t)v_2(\tau)\} = 2\delta(t-\tau)$$

$$Cov\{v_1(t)v_2(\tau)\} = 0$$

4.4-3   *METHOD OF SOLUTION*   This problem is one of a general class of
problems known as <u>Wiener filtering</u> problems [1].   The estimate of $y(t)$ can
be generated by using $z_1(t)$ and $z_2(t)$ as inputs to a linear time invariant
system with transfer functions $G_1(s) = \hat{Y}(s)/Z_1(s)$, $G_2(s) = \hat{Y}(s)/Z_2(s)$.   The
calculation of G_1 and G_2 by standard spectral factorization techniques as
described in[1] is at best difficult and often impossible.   The problem
can be solved, fortunately, by use of a time invariant Kalman filter which
can be determined by using RICATI.

[1] Sage, A. P. and J. L. Melsa, <u>Estimation Theory with Applications to Communi-
cations and Control</u>, McGraw-Hill, New York, 1971.

We first obtain a state variable model for the linear system generating y_1 and y_2. Let x_1, x_2 and x_3 denote the state variables corresponding to the signal model transfer function, x_4 the state variable of the upper channel and x_5 the state variable of the lower channel. Then it can be verified that the following is a valid state variable representation of the system.

$$\dot{x} = Ax + bw$$

$$y_1 = c_1^T x \qquad \text{or} \qquad y = \begin{bmatrix} y_1 \\ y_2 \end{bmatrix} = \begin{bmatrix} c_1^T \\ c_2^T \end{bmatrix} x = Cx$$

$$y_2 = c_2^T x$$

where

$$A = \begin{bmatrix} 0 & 1 & 0 & 0 & 0 \\ 0 & 0 & 1 & 0 & 0 \\ 0 & -6 & -5 & 0 & 0 \\ 1 & 4 & 0 & -1 & 0 \\ 1 & 4 & 0 & 0 & -2 \end{bmatrix}$$

$$b^T = [\, 0 \quad 0 \quad 1 \quad 0 \quad 0 \,]$$

$$c_1^T = [\, 0 \quad 0 \quad 0 \quad 1 \quad 0 \,]$$

$$c_2^T = [\, 0 \quad 0 \quad 0 \quad 0 \quad 1]$$

The required noise covariance matrices are

$$\bar{R} = \begin{bmatrix} 1 & 0 \\ 0 & 2 \end{bmatrix} \qquad \bar{Q} = 1$$

Using the input deck of Fig. 4.4-3 as the input to RICATI, the following steady-state gain was obtained:

$$K_f = \begin{bmatrix} .1332 & .0328 & -.0327 & .2330 & .1317 \\ .0350 & .0121 & -.0106 & .0659 & .0391 \end{bmatrix}$$

Thus the filter which generates estimates of $y(t)$ is given by

$$\dot{\hat{x}}(t) = A\hat{x}(t) + K_f(z(t) - C\hat{x}(t))$$

$$\hat{y}(t) = \hat{x}_1(t) + 4\hat{x}_2(t)$$

140

Fig. 4.4-3

Input Deck for RICATI

Card No.	Column No. 5	10	15	20	25	30	35	40	45
1	COMMUNICATION PROB. 5 / / 2								
2	0.0	1.0		0.0		0.0		0.0	
3	0.0	0.0		1.0		0.0		0.0	
4	0.0	-6.0		-5.0		0.0		0.0	
5	1.0	4.0		0.0		-1.0		0.0	
6	1.0	4.0		0.0		0.0		-2.0	
7	0.0	0.0		1.0		0.0		0.0	
8	0.0	0.0		0.0		1.0		0.0	
9	0.0	0.0		0.0		0.0		1.0	
10	F								
11	1.0	0.0							
12	0.0	2.0							
13	1.0								

A   SUBPROGRAMS

In this appendix all of the subprograms used in the preceding chapters are described with the exception of the plot routines which are discussed in Appendix B. This discussion of subprograms is presented to assist in the understanding of the primary program operations and also to provide a collection of subprograms which may be used to write programs. A cross-listing of subprograms and main programs is presented in Table A-1 for reference. From this table one may see which subprograms are used in any given programs and also which programs use any given subprograms. In addition, page numbers are given so that Table A-1 may also be used as a table of contents for this subprogram listing.

Table A-1

Subprogram - Program Cross List

		BASMAT 2.2	RTRESP 2.3	GTRESP 2.4	SENSIT 2.5	STRVARFDBK 2.6	OBSERV 2.7	LUEN 2.8	SERCOM 2.9	RICATI 2.10	MIMO 2.11	FRESP 3.2	RTLOC 3.3	PRREXP 3.4	PAGE NUMBER
A.1	CALCU			●											144
A.2	CHREQ	●			●										145
A.3	DET	●									●				147
A.4	DIVP													●	148
A.5	FORM				●										150
A.6	MAXI											●			151
A.7	MPY				●										152
A.8	NORMP													●	153
A.9	PADD													●	154
A.10	PEXCG													●	155
A.11	PFEXP													●	156
A.12	PHNOM											●			159
A.13	PMUL													●	160
A.14	PROOT	●	●		●	●		●			●	●	●	●	161
A.15	PVAL											●		●	164
A.16	RUNGE			●											165
A.17	SEMBL		●					●			●	●	●	●	167
A.18	SIMEQ	●				●				●	●				169
A.19	SORT													●	171
A.20	STMST	●	●												172
A.21	SUBP													●	175
A.22	TRESP			●											176
A.23	YDOT			●											179
A.24	LINEQ							●	●						180
A.25	CHREQA		●			●					●				183
A.26	MULT						●				●				185
A.27	HERMIT						●				●				186

A.1 SUBROUTINE CALCU

The subroutine CALCU is one of a package of four subroutines (CALCU, RUNGE, TRESP, and YDOT) used to compute the time response of a closed-loop linear system by the use of a fourth-order Runge-Kutta integration procedure. The CALCU subroutine is used to determine the reference input r(t) and the control input $u(t) = K[r(t) - \underset{\sim}{k}^T \underset{\sim}{x}(t)]$.

A.1-1 *CALLING SEQUENCE*

Call CALCU(X, U, T, N, AK, GAIN, R)

A.1-2 *DEFINITION OF SYMBOLS*

X = State vector array, $\underset{\sim}{x}$(t)
U = Control input, u(t)
T = Time
N = System order, $N \leq 10$
AK = Feedback coefficient array, $\underset{\sim}{k}$
GAIN = Controller gain K
R = Reference input r(t)

A.1-3 *NOTES*

(1) The input function r(t) is defined by inserting the appropriate FORTRAN coding into the subroutine CALCU as indicated. For example the coding shown in the listing below defines the input r(t) $t \geq 0$ given by

$r(t) = 5 \qquad 0 \leq t \leq 1$

$\quad\quad = 0 \qquad$ otherwise

A.1-4 *SUBPROGRAMS USED* None.

A.1-5 *LISTING*

```
      SUBROUTINE CALCU(X,U,T,N,AK,GAIN,R)
C     THIS SUBROUTINE COMPUTES THE REFERENCE AND CONTROL INPUTS
      DIMENSION X(10), AK(10)
C     THE CARDS BETWEEN HERE AND THE NEXT COMMENT CARD
C     ARE SUPPLIED BY THE USER TO DEFINE R(T)
 1000 IF(T.GT.1.0) GO TO 1001
      R =5.0
      GO TO 1002
 1001 R=0.0
 1002 CONTINUE
C     END OF ROUTINE TO DEFINE R(T)
      U = R
      DO 1 I = 1, N
    1 U = U - AK(I)*X(I)
      U = U*GAIN
      RETURN
      END
```

A.2 SUBROUTINE CHREQ

This subroutine is used to determine the characteristic polynomial $\det(sI - A)$ and the resolvent matrix $(sI - A)^{-1}$ for the matrix A.

A.2-1 *CALLING SEQUENCE*

Call CHREQ(A, N, C, NRM)

A.2-2 *DEFINITION OF SYMBOLS*

A = Matrix array
N = Dimension of A, N $\leq$ 10
C = Coefficient array for the characteristic polynomial, C(1) is the coefficient of the constant term
NRM = 0 the matrix coefficients of the resolvent matrix are printed
$\neq$ 0 the matrix coefficients of the resolvent matrix are not printed.

A.2-3 *NOTES*

(1) This subroutine uses the Leverrier algorithm to compute the resolvent matrix $\Phi(s) = (sI - A)^{-1}$ in the form

$$\Phi(s) = \frac{\sum\limits_{i=1}^{N} R_i s^{N-i}}{D(s)}$$

where D(s) is the characteristic polynomial $\det(sI - A)$. Although the algorithm computes both the matrix coefficients R_i and the scalar coefficients of the characteristic polynomial only the latter are returned in the argument list. The matrix coefficients however are stored in the array ZED(I, J, K) = R_I(J, K). See L. A. Zadeh and C. A. Desoer: Linear System Theory, McGraw-Hill Book Co., Inc., 1963, pp. 303-305 for a discussion of the Leverrier algorithm.

(2) The coefficients of the characteristic polynomial are computed by the use of CHREQA.

A.2-4 *SUBPROGRAM USED*

(1) CHREQA

A.2-5 *LISTING*

```
      SUBROUTINE CHREQ(A,N,C,NRM)
C  THIS SUBROUTINE FINDS THE COEFFICIENTS OF THE CHARACTERISTIC
C     POLYNOMIAL USING THE LEVERRIER ALGORITHM
      COMMON ZED(10,10,10)
      DIMENSION A(10,10),C(11),ATEMP(10,10),PROD(10,10)
      DATA ATEMP/100*0.0/
 1000 FORMAT (1H0,5X,31HTHE MATRIX COEFFICIENTS OF THE    ,
     *   33HNUMERATOR OF THE RESOLVENT MATRIX  )
 1001 FORMAT (1H0,  5X,29HTHE MATRIX COEFFICIENT OF S**,I1/)
 1002 FORMAT (1P6E20.7)
 1003 FORMAT (1H0,45(1H*))
      CALL CHREQA(A,N,C)
      DO 65 I=1,N
   65 ATEMP(I,I)=1.0
   70 DO 80 I=1,N
      DO 80 J=1,N
   80 ZED(N,I,J)=ATEMP(I,J)
```

145

```
      IF (NRM.NE.0) GO TO 71
      WRITE (6,1003)
      WRITE (6,1000)
      M=N-1
      WRITE (6,1001) M
      DO 35 I=1,N
35    WRITE (6,1002) (ATEMP(I,J),J=1,N)
71    DO 40 I=1,N
      DO 40 J=1,N
40    ATEMP(I,J)=A(I,J)
      DO 10 I=1,N
      NNN=N-I
      IF (I.EQ.1) GO TO 55
      IF (NRM.NE.0) GO TO 60
      WRITE (6,1001) NNN
      DO 45 J=1,N
45    WRITE (6,1002) (ATEMP(J,K),K=1,N)
60    NP=NNN+1
      DO 90 II=1,N
      DO 90 J=1,N
90    ZED(NP,II,J)=ATEMP(II,J)
      DO 15 J=1,N
      DO 15 K=1,N
      PROD(J,K)=0.0
      DO 15 L=1,N
15    PROD(J,K)=PROD(J,K)+(A(J,L)*ATEMP(L,K))
      DO 13 J=1,N
      DO 13 K=1,N
13    ATEMP(J,K)=PROD(J,K)
55    DO 10 J=1,N
10    ATEMP(J,J)=ATEMP(J,J)+C(N-I+1)
      RETURN
      END
```

A.3 FUNCTION DET

The function DET is used to compute the determinant of a matrix.

A.3-1 *CALLING SEQUENCE*

X = DET(A, KC)

A.3-2 *DEFINITION OF SYMBOLS*

A = Matrix array, i.e. A(I, J) = the element of the matrix in the I-th row and J-th column.

KC = Order of the matrix, KC ≤ 10

A.3-3 *NOTES*

(1) If the matrix A is singular, the program will simply set DET = 0.

(2) This subprogram uses a Gauss elimination method to place the matrix in upper triangular form. See R. W. Hamming: <u>Numerical Methods for Scientists and Engineers</u>, McGraw-Hill Book Company, Inc., 1962, pp. 360-366.

A.3-4 *SUBPROGRAMS USED* None.

A.3-5 *LISTING*

```
      FUNCTION DET(A,KC)
C  THIS FUNCTION SUBPROGRAM FINDS THE DETERMINANT OF A MATRIX
C     USING DIAGONALIZATION PROCEDURE
      DIMENSION A(10,10),B(10,10)
      IREV =0
      DO1 I=1,KC
      DO1 J=1,KC
    1 B(I,J)=A(I,J)
      DO20I=1,KC
      K=I
    9 IF(B(K,I))10,11,10
   11 K=K+1
      IF(K-KC) 9,9,51
   10 IF(I-K) 12,14,51
   12 DO13M=1,KC
      TEMP=B(I,M)
      B(I,M)=B(K,M)
   13 B(K,M)=TEMP
      IREV = IREV+1
   14 II=I+1
      IF(II.GT.KC) GO TO 20
      DO17 M=II,KC
   18 IF(B(M,I)) 19,17,19
   19 TEMP =B(M,I)/B(I,I)
      DO16N=I,KC
   16 B(M,N)=B(M,N)-B(I,N)*TEMP
   17 CONTINUE
   20 CONTINUE
      DET=1.
      DO2 I=1,KC
    2 DET=DET*B(I,I)
      DET=(-1.)**IREV*DET
      RETURN
   51 DET=0
      RETURN
      END
```

A.4 SUBROUTINE DIVP

The subroutine DIVP is used to form the quotient of two polynomials as

$$P(s) = \frac{X(s)}{Y(s)}$$

A.4-1 *CALLING SEQUENCE*

CALL DIVP (P, IP, X, IXA, Y, IYA, TOL, IER)

A.4-2 *DEFINITION OF SYMBOLS*

P = Coefficient array for P(s), P(1) is the constant term
IP = Order of P(s)
X = Coefficient array for X(s)
IXA = Order of X(s)
Y = Coefficient array for Y(s)
IYA = Order of Y(s)
TOL = Tolerance value
IER = Error indicator = 0 no error
                      = 1 error

A.4-3 *NOTES*

(1) This subroutine is the subroutine NORMP to discard coefficients
starting with the highest order until one which exceeds TOL is found. This is
done of Y(s) before division and on P(s) after division. If Y(s) has no terms
which exceed TOL, then IER is set to 1.

A.4-4 *SUBPROGRAMS USED*

(1) NORMP

A.4-5 *LISTING*

```
      SUBROUTINE DIVP(P,IP,X,IXA,Y,IYA,TOL,IER)
C     THIS SUBROUTINE DIVIDES TWO POLYNOMIALS
      DIMENSION X(10), Y(10), P(10)
      IX=IXA+1
      IY=IYA+1
      CALL NORMP (Y,IY,TOL)
      IF (IY) 50,50,10
10    IP = IX - IY +1
      IF (IP) 20,30,60
20    IP = 0
30    IER = 0
40    RETURN
50    IER = 1
      GO TO 40
60    IX = IY -1
      I = IP
70    II = I + IX
      P(I) = X(II)/Y(IY)
      DO 80 K =1, IX
      J = K-1+I
      X(J) = X(J) -P(I)*Y(K)
80    CONTINUE
      I = I-1
      IF (I) 90,90,70
90    CALL NORMP(X,IX,TOL)
```

148

```
IP=IP-1
GO TO 30
END
```

A.5   SUBROUTINE FORM

The subroutine FORM is used to compute the weighted sum of two polynomials A(s) and B(s) as

$$C(s) = B(s) + GA(s)$$

and to determine the order of the resulting polynomial C(s).

A.5-1   *CALLING SEQUENCE*

CALL FORM (G, A, N, B, M, C, IX)

A.5-2   *DEFINITION OF SYMBOLS*

G = Scalar weighting term
A = Polynomial coefficient array for A(s), constant term first
N = Order of the polynomial A(s), $N \leq 10$
B = Polynomial coefficient array for B(s), constant term first
M = Order of the polynomial B(s), $M \leq 10$
C = Polynomial coefficient array for the resulting polynomial C(s)
IX = Order of the polynomial C(s)

A.5-3   *SUBPROGRAMS USED*   None.

A.5-4   *LISTING*

```
      SUBROUTINE FORM(G,A,N,B,M,C,IX)
C   THIS SUBROUTINE FORMS THE WEIGHTED SUM OF TWO POLYNOMIALS
      DIMENSION A(11),B(11),C(11)
      IF(N-M) 1,2,2
    1 IX=M+1
      GO TO 3
    2 IX=N+1
    3 DO 4 I=1,IX
    4 C(I)=B(I)+G*A(I)
      IX=IX-1
      RETURN
      END
```

A.6   SUBROUTINE MAXI

The subroutine MAXI does a compare and replace search of the array AT against a maximum value AMX and a minimum value AMN.

A.6-1   *CALLING SEQUENCE*

CALL MAXI (AT, ND, AMX, AMN)

A.6-2   *DEFINITION OF SYMBOLS*

AT = Array to be searched
ND = Number of elements of AT to be searched
AMX = Maximum value
AMN = Minimum value

A.6-3   *NOTES*

(1) This subroutine searches the first ND elements of the array AT against the values AMX and AMN. If the minimum elements of AT is less than AMN, that value replaces AMN. Similarly if the maximum valued element of AT exceeds AMX, then that value replaces AMX.

(2) This subroutine may be used to search the array AT for its minimum and maximum values by setting AMX = AT(1) and AMN = AT(1) before the CALL statement.

A.6-4   *SUBPROGRAMS USED*   None.

A.6-5   *LISTING*

```
      SUBROUTINE MAXI(AT,ND,AMX,AMN)
C     THIS IS A COMPARE AND REPLACE ROUTINE
      DIMENSION AT(ND)
      DO 2 J = 1,ND
      IF(AT(J) - AMN) 8,9,9
    8 AMN = AT(J)
    9 IF(AT(J) - AMX) 2,2,7
    7 AMX - AT(J)
    2 CONTINUE
      RETURN
      END
```

A.7 SUBROUTINE MPY

The subroutine MPY is used to compute a scalar polynomial A(s) from a N by N matrix polynomial $\underset{\sim}{Z}(s)$ in the form

$$A(s) = \underset{\sim}{g}^T \underset{\sim}{Z}(s) \underset{\sim}{b}$$

where $\underset{\sim}{g}$ and $\underset{\sim}{b}$ are constant vectors.

A.7-1 *CALLING SEQUENCE*

CALL MPY(B, G, N, ANS, NS)

A.7-2 *DEFINITION OF SYMBOLS*

B = Vector array for $\underset{\sim}{b}$
G = Vector array for $\underset{\sim}{g}$
N = Dimension of $\underset{\sim}{b}$ and $\underset{\sim}{g}$, $N \leq 10$
ANS = Polynomial coefficient array for A(s)
NS = Order of the polynomial A(s).

A.7-3 *NOTES*

(1) The matrix polynomial $\underset{\sim}{Z}(s)$ takes the form of matrix coefficients of powers of s such as

$$\underset{\sim}{Z}(s) = \underset{\sim}{\Psi}_1 + \underset{\sim}{\Psi}_2 s + \ldots + \underset{\sim}{\Psi}_N s^{N-1}$$

Note that $\underset{\sim}{Z}(s)$ is a polynomial of order N-1. The matrix coefficient $\underset{\sim}{\Psi}_i$ are entered into the subroutine in the array ZED(I, J, K) by the use of a common statement with ZED(I, J, K) = $[\Psi_i]_{jk}$.

A.7-4 *SUBPROGRAMS USED* None.

A.7-5 *LISTING*

```
      SUBROUTINE MPY(B,G,N,ANS,NS)
C   FORMS A SCALAR POLYNOMIAL FROM A MATRIX POLYNOMIAL
      COMMON ZED(10,10,10)
      DIMENSION B(10),G(10),W(10,10),ANS(11)
      DO 1  I=1,N
      DO 1  J=1,N
      W(J,I)=0.0
      DO 1  K=1,N
    1 W(J,I)=W(J,I)+ZED(I,J,K)*B(K)
      DO 2  I=1,N
      ANS(I)=0.0
      DO 2  J=1,N
    2 ANS(I)=ANS(I)+W(J,I)*G(J)
      NN=N+1
      ANS(NN)=0.0
      NS=N-1
      RETURN
      END
```

A.8 SUBROUTINE NORMP

The subroutine NORMP is used to check the coefficients of a polynomial starting the coefficient of the highest order and set them to zero if they do not exceed a value EPS until one which does exceed EPS is found.

A.8-1 *CALLING SEQUENCE*

    CALL NORMP (X, IX, EPS)

A.8-2 *DEFINITION OF SYMBOLS*

    X = Coefficient array, X(1) is the constant term
    IX = Order of the polynomial +1
    EPS = Threshold of the search

A.8-3 *NOTES*

    (1) This subroutine is used in connection with DIVP to eliminate extraneous terms

A.8-4 *SUBPROGRAMS USED*   None.

A.8-5 *LISTING*

```
      SUBROUTINE NORMP(X,IX,EPS)
C     THIS SUBROUTINE ZEROES COEFFICIENTS OF A POLYNOMIAL
      DIMENSION X(10)
1 IF (IX) 4,4,2
2 IF(ABS(X(IX))-EPS) 3,3,4
3 IX = IX -1
      GO TO 1
4 RETURN
      END
```

A.9   SUBROUTINE PADD

The subroutine PADD is used to add two polynomials as

$$Z(s) = X(s) + Y(s)$$

A.9-1   *CALLING SEQUENCE*

CALL PADD (Z, IZ, X, IXA, Y, IYA)

A.9-2   *DEFINITION OF SYMBOLS*

Z = Coefficient array for $Z(s)$, $Z(1)$ is the constant term
IZ = Order of $Z(s) \leq 9$
X = Coefficient array for $X(s)$
IXA = Order of $X(s)$
Y = Coefficient array for $Y(s)$
IYA = Order of $Y(s)$

A.9-3   *SUBPROGRAMS USED*   None.

A.9-4   *LISTING*

```
      SUBROUTINE PADD(Z,IZ,X,IXA,Y,IYA)
C     THIS SUBROUTINE ADDS TWO POLYNOMIALS
      DIMENSION Z(10), X(10), Y(10)
      IX=IXA+1
      IY=IYA+1
      ND = IX
      IF (IX-IY) 10,20,20
   10 ND = IY
   20 IF (ND) 90,90,30
   30 DO 80 I = 1, ND
      IF (I-IX) 40,40,60
   40 IF(I-IY) 50,50,70
   50 Z(I) = X(I) + Y(I)
      GO TO 80
   60 Z(I) = Y(I)
      GO TO 80
   70 Z(I) = X(I)
   80 CONTINUE
   90 IZ=ND-1
      RETURN
      END
```

A.10 SUBROUTINE PEXCG

The subroutine is used to replace the array A with the array B; the array A is destroyed.

A.10-1 *CALLING SEQUENCE*

    CALL PEXCG (A, IA, B, IB)

A.10-2 *DEFINITION OF SYMBOLS*

    A = Array of A
    IA = Order of A
    B = Array for B
    IB = Order of B

A.10-3 *NOTES*

    (1) Upon **exit** the first IB elements of the array A will be equal to the first IB elements of B and IA = IB.

A.10-4 *SUBPROGRAMS USED* None.

A.10-5 *LISTING*

```
      SUBROUTINE PEXCG(A,IA,B,IB)
C     THIS SUBROUTINE REPLACES A WITH B
      DIMENSION A(1),B(1)
      JJ=IB+1
      DO 1 I=1,JJ
    1 A(I)=B(I)
      IA=IB
      RETURN
      END
```

A.11 SUBROUTINE PFEXP

The subroutine PFEXP is used to obtain the partial fraction expansion of the rational function G(s) given by

$$G(s) = \frac{A(s)}{\displaystyle\prod_{i=1}^{L} (s - s_i)^{m_i}}$$

where L is the number of distinct roots and m_i is the multiplicity of the ith root. The resulting partial fraction expansion takes the form

$$G(s) = \sum_{i=1}^{L} \sum_{j=1}^{m_i} \frac{R_{ij}}{(s - s_i)^j}$$

A.11-1 *CALLING SEQUENCE*

　　　　CALL PFEXP (A, NNM, CR, CI, M, IT, RESR, RESI)

A.11-2 *DEFINITION OF SYMBOLS*

　　　　A = Coefficient array for A(s)
　　　　NNM = Order of the polynomial A(s), ≤ 9
　　　　CR = Real root array
　　　　CI = Imaginary root array
　　　　M = Multiplicity array
　　　　IT = Number of distinct roots
　　　　RESR = Real part of the partial fraction coefficient array R_{ij}
　　　　RESI = Imaginary part of the partial fraction coefficient array R_{ij}

A.11-3 *NOTES*

　　　　(1) This subroutine forms the heart of the PRFEXP program; see Sec. 3.4 for more details on the PFEXP subroutine.

　　　　(2) All complex roots must be simple.

A.11-4 *SUBPROGRAMS USED*

　　　　(1) PADD
　　　　(2) PEXCG
　　　　(3) PMUL
　　　　(4) PVAL
　　　　(5) SUBP

A.11-5 *LISTING*

```
      SUBROUTINE PFEXP(A,NNM,CR,CI,MST,IT,RESR,RESI)
C     THIS SUBROUTINE MAKES A PARTIAL FRACTION EXPANSION
      DIMENSION A(10),CR(10),CI(10),RESR(10,10),RESI(10,10),
     *     D(10),P(10),Q(10),DX(10),DXX(10),M(10),MST(10),
     *     ANS(10)
      DO 198 I=1,IT
  198 M(I)=MST(I)
      DO 199 I=1,10
      D(I)=0.0
      P(I)=0.0
      Q(I)=0.0
      DXX(I)=0.0
  199 DX(I)=0.0
  200 DO 30 I=1, IT
```

```
      CALL PVAL (A, NNM , CR(I), CI(I), VR, VI)
      IMI = M(I)
      PR = 1.
      PI= 0.
      IF (IMI) 30,30,390
390 DO 31 J = 1, IT
      IF (I-J) 32,31,32
 32 IMJ = M(J)
      IF (IMJ) 31,31,33
 33 A1 = CR(I) - CR(J)
      B1 = CI(I) - CI(J)
      DO 34 K = 1, IMJ
      A2 =PR*A1 -PI*B1
      B2 =PR*B1 + A1*PI
      PR = A2
      PI = B2
 34 CONTINUE
 31 CONTINUE
      DIV = PR*PR + PI*PI
      RESR(I,IMI) = (PR*VR + PI*VI)/DIV
      RESI(I,IMI) = (PR*VI - PI*VR)/DIV
 30 CONTINUE
      DO 300 I = 1, IT
      IF (M(I)-1) 300,300 ,301
300 CONTINUE
      GO TO 205
301 JJ = 0
      DO 40 I = 1, 10
      Q(I) = 0.
      P(I) = 0.
 40 CONTINUE
      IQ=0
      IP=0
      Q(1) = 1.
 42 JJ = JJ +1
      ICH = M(JJ)
      IF (ICH) 42,42,422
422 IF (CI(JJ)) 43,45,43
 43 D(1) = -2.*RESI(JJ,ICH)*CI(JJ) -2.*RESR(JJ,ICH)*CR(JJ)
      D(2) = 2.*RESR(JJ,ICH)
      ID=1
      CALL PMUL (DX, IDX, D, ID, Q, IQ)
      D(1) = CI(JJ)*CI(JJ) +CR(JJ)*CR(JJ)
      D(2) = 2.*CR(JJ)
      D(3) = 1.
      ID=2
      CALL PMUL (DXX, IDXX, P, IP, D, ID)
      CALL PADD (P, IP, DX, IDX, DXX, IDXX)
      CALL PMUL (DX, IX, Q, IQ, D, ID)
      CALL PEXCG(Q,IQ,DX,IX)
      JJ = JJ +1
      IF (IT -JJ) 500,500,42
 45 D(1) = -CR(JJ)
      D(2) = 1.
      ID=1
      DO 46 IN = 1, ICH
      CALL PMUL (DX, IXX, D, ID, P, IP)
      CALL PEXCG(P,IP,DX,IXX)
 46 CONTINUE
      IQP1=IQ+1
      DO 48 IXX = 1, IQP1
      DXX(IXX) = RESR(JJ,ICH)*Q(IXX)
 48 CONTINUE
```

```
      IK = IP
      CALL PADD (P, IP, DX, IK, DXX, IQ)
      DO 49 IXX = 1, ICH
      CALL PMUL (DX, IKX, D, ID, Q, IQ)
      CALL PEXCG(Q,IQ,DX,IKX)
   49 CONTINUE
      IQ = IKX
      IF (JJ -IT) 42, 500, 500
  500 CALL SUBP ( ANS, IANS, A, NNM , P, IP)
      CALL PEXCG(A,NNM,ANS,IANS)
      TOL=0.001
      I = 0
  501 I =I+1
      IF(M(I)) 501, 501, 502
  502 M(I) = M(I) - 1
      IF(CI(I)) 507,504,507
  507 D(1) = CR(I)*CR(I) + CI(I)*CI(I)
      D(2) = -2. *CR(I)
      D(3) = 1.
      ID=2
      I = I +1
      M(I) = M(I) - 1
      GO TO 505
  504 D(1) = -CR(I)
      D(2) = 1.
      ID=1
  505 CALL DIVP (ANS,IANS,A,NNM ,D,ID,TOL,IER)
      CALL PEXCG(A,NNM,ANS,IANS)
      IF (I-IT) 501,200,200
  205 RETURN
      END
```

A.12 FUNCTION PHNOM

The function PHNOM is used to normalize an angle X(in degrees) to the range −180° to +180°.

A.12-1 *CALLING SEQUENCE*

Y = PHNOM(X)

A.12-2 *DEFINITION OF SYMBOLS*

X = Angle in degrees

A.12-3 *NOTES*

(1) Either +360° or −360° is added to the angle X until it lies between −180° to +180°.

A.12-4 *SUBPROGRAMS USED*   None.

A.12-5 *LISTING*

```
      FUNCTION PHNOM(X)
C     THIS FUNCTION NORMALIZES AN ANGLE
    7 IF(X − 180.) 4,4,5
    5 X = X − 360.
      GO TO 7
    4 IF(X+180.) 8,8,6
    8 X = X + 360.
      GO TO 4
    6 PHNOM = X
      RETURN
      END
```

A.13 SUBROUTINE PMUL

The subroutine PMUL is used to multiply a polynomial X(s) times Y(s) to obtain Z(s).

A.13-1 *CALLING SEQUENCE*

CALL PMUL (Z, IZ, X, IXA, Y, IYA)

A.13-2 *DEFINITION OF SYMBOLS*

Z = Resulting coefficient array, Z(1) is the constant term
IZ = Order of Z(s), $\leq$ 8
X = Coefficient array for X(s)
IXA = Order of X(s)
Y = Coefficient array for Y(s)
IYA = Order of Y(s)

A.13-3 *SUBPROGRAMS USED*   None.

A.13-4 *LISTING*

```
      SUBROUTINE PMUL(Z,IZ ,X,IXA,Y,IYA)
C     THIS SUBROUTINE FORMS PRODUCT OF TWO POLYNOMIALS
      DIMENSION X(10), Y(10), Z(10)
      IX=IXA+1
      IY=IYA+1
      IF (IX*IY) 10,10,20
   10 IZ = 0
      GO TO 50
   20 IZ=IX+IY
      DO 30 I = 1, IZ
   30 Z(I) = 0.
      DO 40 I = 1, IX
      DO 40 J = 1, IY
      K = I +J -1
      Z(K) = X(I)*Y(J) +Z(K)
   40 CONTINUE
      IZ=IZ-2
   50 RETURN
      END
```

160

A.14 SUBROUTINE PROOT

The subroutine PROOT is used to find the roots of a polynomial with real coefficients.

A.14-1 *CALLING SEQUENCE*

Call PROOT(N, A, U, V, IR)

A.14-2 *DEFINITION OF SYMBOLS*

N = Degree of polynomial, $N \leq 19$
A = Polynomial coefficient array.
U = Real root array.
V = Imaginary root array.
IR = +1 if polynomial is written as

$$A(1) + A(2)s + A(3)s^2 + \ldots + A(N+1)s^N$$

−1 if polynomial is written as
$$A(1)s^N + A(s)s^{N-1} + \ldots + A(n+1)$$

A.14-3 *NOTES*

(1) This subroutine uses a modified Bairstow method for root extraction. See R. W. Hamming: <u>Numerical Methods for Scientists and Engineers</u>, McGraw-Hill Book Company, Inc., 1962, pp. 356-359.

A.14-4 *SUBPROGRAMS USED*   None.

A.14-5 *LISTING*

```
      SUBROUTINEPROOT(N,A,U,V,IR)
C   THIS SUBROUTINE USES A MODIFIED BARSTOW METHOD TO FIND THE
C     ROOTS OF A POLYNOMIAL.
      DIMENSION A(20),U(20),V(20),H(21),B(21),C(21)
      IREV=IR
      NC=N+1
      DO11I=1,NC
    1 H(I)=A(I)
      P=0.
      Q=0.
      R=0.
    3 IF(H(1))4,2,4
    2 NC=NC-1
      V(NC)=0.
      U(NC)=0.
      DO1002I=1,NC
 1002 H(I)=H(I+1)
      GOTO3
    4 IF(NC-1)5,100,5
    5 IF(NC-2)7,6,7
    6 R=-H(1)/H(2)
      GOTO50
    7 IF(NC-3)9,8,9
    8 P=H(2)/H(3)
      Q=H(1)/H(3)
      GOTO70
    9 IF(ABS (H(NC-1)/H(NC))-ABS (H(2)/H(1)))10,19,19
   10 IREV=-IREV
      M=NC/2
      DO11I=1,M
      NL=NC+1-I
```

```
      F=H(NL)
      H(NL)=H(I)
11    H(I)=F
      IF(Q)13,12,13
12    P=0.
      GOTO15
13    P=P/Q
      Q=1./Q
15    IF(R)16,19,16
16    R=1./R
19    E=5.E-10
      B(NC)=H(NC)
      C(NC)=H(NC)
      B(NC+1)=0.
      C(NC+1)=0.
      NP=NC-1
20    DO49J=1,1000
      DO21I1=1,NP
      I=NC-I1
      B(I)=H(I)+R*B(I+1)
21    C(I)=B(I)+R*C(I+1)
      IF(ABS (B(1)/H(1))-E)50,50,24
24    IF(C(2))23,22,23
22    R=R+1.
      GOTO30
23    R=R-B(1)/C(2)
30    DO37I1=1,NP
      I=NC-I1
      B(I)=H(I)-P*B(I+1)-Q*B(I+2)
37    C(I)=B(I)-P*C(I+1)-Q*C(I+2)
      IF(H(2))32,31,32
31    IF(ABS (B(2)/H(1))-E)33,33,34
32    IF(ABS (B(2)/H(2))-E)33,33,34
33    IF(ABS (B(1)/H(1))-E)70,70,34
34    CBAR=C(2)-B(2)
      D=C(3)**2-CBAR*C(4)
      IF(D)36,35,36
35    P=P-2.
      Q=Q*(Q+1.)
      GOTO49
36    P=P+(B(2)*C(3)-B(1)*C(4))/D
      Q=Q+(-B(2)*CBAR+B(1)*C(3))/D
49    CONTINUE
      E=E*10.
      GOTO20
50    NC=NC-1
      V(NC)=0.
      IF(IREV)51,52,52
51    U(NC)=1./R
      GOTO53
52    U(NC)=R
53    DO54I=1,NC
54    H(I)=B(I+1)
      GOTO4
70    NC=NC-2
      IF(IREV)71,72,72
71    QP=1./Q
      PP=P/(Q*2.0)
      GOTO73
72    QP=Q
      PP=P/2.0
73    F=(PP)**2-QP
      IF(F)74,75,75
```

162

```
 74 U(NC+1)=-PP
    U(NC)=-PP
    V(NC+1)=SQRT (-F)
    V(NC)=-V(NC+1)
    GOTO76
 75 IF(PP)81,80,81
 80 U(NC+1)=-SQRT(F)
    GO TO 82
 81 U(NC+1)=-(PP/ABS (PP))*(ABS (PP)+SQRT (F))
 82 CONTINUE
    V(NC+1)=0.
    U(NC)=QP/U(NC+1)
    V(NC)=0.
 76 DO77I=1,NC
 77 H(I)=B(I+2)
    GOTO4
100 RETURN
    END
```

A.15 SUBROUTINE PVAL

The subroutine PVAL is used to evaluate a polynomial A(s) for s = PR + jRI.
The coefficients of A(s) are real.

A.15-1 *CALLING SEQUENCE*

    CALL PVAL (A, NN, PR, PI, VR, VI)

A.15-2 *DEFINITION OF SYMBOLS*

    A = Polynomial array, A(1) is the constant term
    NN = Order of A(s)
    PR = Real value of s
    PI = Imaginary value of s
    VR = Real value of A(s)
    VI = Imaginary value of A(s)

A.15-3 *SUBPROGRAMS USED* None.

A.15-4 *LISTING*

```
      SUBROUTINE PVAL(A,NN,PR,PI,VR,VI)
C     USED TO DETERMINE THE VALUE OF A POLYNOMIAL
      DIMENSION A(20)
      COMPLEX S,P
      S=CMPLX(PR,PI)
      P=CMPLX(A(NN+1),0.)
      DO 100 J=1,NN
  100 P=P*S+A(NN+1-J)
      VR=REAL(P)
      VI=AIMAG(P)
      RETURN
      END
```

A.16 SUBROUTINE RUNGE

The subroutine RUNGE is one of a package of four subroutines (CALCU, RUNGE, TRESP and YDOT) used to compute the time response of a closed-loop linear system by the use of a fourth order Runge-Kutta integration procedure. The RUNGE subroutine contains the actual Runge-Kutta integration algorithm.

A.16-1 *CALLING SEQUENCE*

```
      I = 0
1000 CALL RUNGE (N, FN, U, X, Y, L, I)
      IF (L.EQ.2) GO TO 2000
      CALL CALCU (Y, U, N, AK, GAIN, R)
      CALL YDOT (A, Y, FN, B, U, N)
      GO TO 1000
2000 CONTINUE
```

A.16-2 *DEFINITION OF SYMBOLS*

N = System order, $N \leq 10$
FN = Derivative array, $x(t)$, from YDOT
H = Integration step size
X = Time variable, t
Y = State vector array, $x(t)$
L = 1 indicates that the integration process is incomplete
  = 2 indicates that an integration step has been completed
I = Number of times RUNGE has been entered on this integration step.

A.16-3 *NOTES*

(1) The subroutine RUNGE is entered four times for each integration step. At the second and fourth entrances, the variable X is incremented by 0.5H.

(2) The counting variable I must be zero before RUNGE is initially; after that I is automatically reset to zero after each complete integration step.

(3) See R. W. Hamming: <u>Numerical Methods for Scientists and Engineers</u>, McGraw-Hill Book Company, Inc., 1962, pp. 211-214 for a discussion of the Runge-Kutta integration algorithm.

A.16-4 *SUBPROGRAMS USED* None.

A.16-5 *LISTING*

```
      SUBROUTINE RUNGE (N,FN, H, X, Y, L,I)
C     FOURTH ORDER RUNGE KUTTA INTEGRATION ROUTINE
      DIMENSION Y(600), SAVEY(600), PHI(600), FN(8)
      I = I + 1
      GO TO ( 1, 2, 3, 4, 5) , I
    1 L = 1
      RETURN
    2 DO 600 J=1,N
      SAVEY(J) = Y(J)
      PHI(J) = FN(J)
  600 Y(J) = SAVEY(J) + .5*H*FN(J)
      X = X + .5*H
      L = 1
      RETURN
    3 DO 700 J=1,N
      PHI(J) = PHI(J) + 2.*FN(J)
  700 Y(J) = SAVEY(J) + .5*H*FN(J)
```

```
      L = 1
      RETURN
   4  DO 800 J=1,N
      PHI(J) = PHI(J) + 2.*FN(J)
800   Y(J) = SAVEY(J) + H*FN(J)
      X = X + .5*H
      L = 1
      RETURN
   5  DO 900 J=1,N
900   Y(J) = SAVEY(J) + (H/6.)*(PHI(J) + FN(J))
      L = 2
      I = 0
      RETURN
      END
```

A.17 SUBROUTINE SEMBL

This subroutine is used to determine the coefficients of a polynomial from the roots of the polynomial.

A.17-1 *CALLING SEQUENCE*

Call SEMBL (N, RR, RI, CF)

A.17-2 *DEFINITION OF SYMBOLS*

N = Polynomial order, N $\leq$ 10
RR = Real root array
RI = Imaginary root array
CF = Polynomial coefficient array in the form CF(1) + CF(s)s + ... CF(N+1)s^N. The coefficient CF(N+1) is set to unity.

A.17-3 *NOTES*

(1) This subroutine computes the polynomial coefficients by the use of the algorithm

$$CF(I) = (-1)^{N-I+1} \Sigma \text{ (product of the roots taken N-I+1 at a time)}$$
$$I = 1, 2, ..., N$$

(2) Complex conjugate roots are handled directly by the use of complex arithmetic and an error message is printed if the coefficient array has any complex values.

A.17-4 *SUBPROGRAMS USED*   None.

A.17-5 *LISTING*

```
      SUBROUTINE SEMBL(N,RR,RI,CF)
C THIS SUBROUTINE DETERMINES THE COEFFICIENTS OF A
C POLYNOMIAL FROM ITS ROOTS
      COMPLEX R(10),C(11),PR,SUM
      DIMENSION RR(10),RI(10),J(11),CF(11)
      NN=N+1
      DO 10 I=1,N
   10 R(I)=CMPLX(RR(I),RI(I))
      CF(NN)=1.0
      DO 14 M=1,N
      SUM=CMPLX(0.0,0.0)
      L=1
      J(1)=1
      GO TO 2
    1 J(L)=J(L)+1
    2 IF(L-M) 3,5,50
    3 MM=M-1
      DO 4 I=L,MM
      II=I+1
    4 J(II)=J(I)+1
    5 PR=CMPLX(1.0,0.0)
      DO 7 I=1,M
      ICK=J(I)
    7 PR=-PR*R(ICK)
      SUM=SUM+PR
      DO 6 I=1,M
      L=M-I+1
      IF(J(L)-N+M-L) 1,6,50
    6 CONTINUE
      MP=N-M+1
```

167

```
  14 CF(MP)=REAL(SUM)
     RETURN
  50 PRINT 2000
2000 FORMAT (/,4X,14HERROR IN SEMBL)
     RETURN
     END
```

A.18 SUBROUTINE SIMEQ

The subroutine SIMEQ is used to solve a set of simultaneous linear
equations of the form $\underline{A}\underline{X} = \underline{XDOT}$ or to invert a matrix. Here the square matrix
$\underline{A}$ and the column vector $\underline{XDOT}$ are given and the solution $\underline{X} = \underline{A}^{-1}(\underline{XDOT})$ is
desired.

A.18-1 *CALLING SEQUENCE*

Call SIMEQ(A, XDOT, KC, AINV, X, IERR)

A.18-2 *DEFINITION OF SYMBOLS*

A = Coefficient array matrix
XDOT = Coefficient vector
KC = Order of the matrix, KC $\leq$ 10
AINV = Inverse of the matrix $\overline{A}$
X = Solution of the simultaneous equation, i.e. X = (AINV)XDOT
IERR = 1 if A is nonsingular
     = 0 if A is singular

A.18-3 *NOTES*

(1) This subroutine uses a Gauss elimination method with column
pivoting to determine AINV and X. See R. W. Hamming: <u>Numerical Methods for
Scientists and Engineers</u>, McGraw-Hill Book Company, Inc., 1962, pp. 360-366.

(2) The subroutine prints the error message: "THE MATRIX IS SINGULAR"
if the matrix A is singular. In addition the variable IERR is set to zero.

(3) If the subroutine is used to invert a matrix the arguments X
and XDOT may be assigned to any temporary variables. X will be destroyed but
XDOT will be unchanged.

A.18-4 *SUBPROGRAMS USED* None.

A.18-5 *LISTING*

```
      SUBROUTINE SIMEQ (A,XDOT,KC,AINV,X,IERR)
C   THIS SUBROUTINE FINDS THE INVERSE OF THE MATRIX A USING
C     DIAGONALIZATION PROCEDURES.
      DIMENSION A(10,10),B(10,10),XDOT(11),X(11),AINV(10,10)
      N=1
      IERR=1
      DO1 I=1,KC
      DO1 J=1,KC
      AINV(I,J)=0
    1 B(I,J)=A(I,J)
      DO2 I=1,KC
      AINV(I,I)=1
    2 X(I)=XDOT(I)
      DO3 I=1,KC
      COMP=0
      K=I
    6 IF(ABS (B(K,I))-ABS (COMP))5,5,4
    4 COMP=B(K,I)
      N=K
    5 K=K+1
      IF(K-KC)6,6,7
    7 IF(B(N,I))8,51,8
    8 IF(N-I)51,12,9
    9 DO10 M=1,KC
      TEMP=B(I,M)
      B(I,M)=B(N,M)
```

169

```
      B(N,M)=TEMP
      TEMP=AINV(I,M)
      AINV(I,M)=AINV(N,M)
10    AINV(N,M)=TEMP
      TEMP=X(I)
      X(I)=X(N)
      X(N)=TEMP
12    X(I)=X(I)/B(I,I)
      TEMP = B(I,I)
      DO13 M=1,KC
      AINV(I,M) = AINV(I,M)/TEMP
13    B(I,M) = B(I,M)/TEMP
      DO16 J=1,KC
      IF(J-I)14,16,14
14    IF(B(J,I))15,16,15
15    X(J)=X(J)-B(J,I)*X(I)
      TEMP=B(J,I)
      DO17 N=1,KC
      AINV(J,N)=AINV(J,N)-TEMP*AINV(I,N)
17    B(J,N)=B(J,N)-TEMP*B(I,N)
16    CONTINUE
 3    CONTINUE
      RETURN
51    PRINT 52
52    FORMAT (     6X,22HTHE MATRIX IS SINGULAR)
      IERR=0
      RETURN
      END
```

A.19 SUBROUTINE SORT

The subroutine SORT is used to sort the elements of the arrays AR, AI and ICK and places the elements in increasing value.

A.19-1 *CALLING SEQUENCE*

CALL SORT (AR, AI, IT, ICK)

A.19-2 *DEFINITION OF SYMBOLS*

AR = Primary array
AI = Secondary array
IT = Dimension of arrays, IT $\leq$ 10
ICK = Additional array

A.19-3 *NOTES*

(1) The elements in the three arrays AR, AI and ICK which have the same index before the sort will have the same index after the sort. In a sense, these three arrays are "characters" in a "word" and the sort is to arrange words. The sort begins with the elements of AR. If AR(I) is less than AR(J), J < I, then AR(I) and AI(I) and ICK(I) replace AR(J), AI(J) and ICK(J). If AR(I) = AR(J), then the sort is made on the absolute value of AI(I) and AI(J).

A.19-4 *SUBPROGRAMS USED*   None.

A.19-5 *LISTING*

```
      SUBROUTINE SORT(AR,AI,IT, ICK)
C     THIS SUBROUTINE SORTS THE ARRAYS AR, AI, ICK
      DIMENSION AR(10), AI(10), ICK(10)
      ITM1 = IT -1
      IF(ITM1) 30,30,1
    1 DO 10 I = 1, ITM1
      IP1 = I +1
      DO 20 J = IP1, IT
      IF(AR(I)-AR(J)) 20,2,3
    2 IF(ABS(AI(I))-ABS(AI(J))) 20,20,3
    3 SAVE1 = AR(I)
      SAVE2 = AI(I)
      ISAVE = ICK(I)
      AR(I) = AR(J)
      AI(I) = AI(J)
      ICK(I) = ICK(J)
      AR(J) = SAVE1
      AI(J) = SAVE2
      ICK(J) = ISAVE
   20 CONTINUE
   10 CONTINUE
   30 RETURN
      END
```

A.20   SUBROUTINE STMST

The subroutine STMST is used to compute the state transition matrix $\Phi(t) = \exp\{\underset{\sim}{A}t\}$ for a matrix $\underset{\sim}{A}$.

A.20-1   *CALLING SEQUENCE*

   Call STMST (N, A, EIGR, EIGI, IKNOW)

A.20-2   *DEFINITION OF SYMBOLS*

   N = Order of the matrix A, N $\leq$ 10
   A = Matrix array
   EIGR = Real-part-of-the-eigenvalue array
   EIGI = Imaginary-part-of-the-eigenvalue array
   IKNOW = 0 state transition matrix is printed
          $\neq$ 0 state transition matrix is not printed

A.20-3   *NOTES*

   (1)   This subroutine uses the Sylvester Expansion Theorem to determine the state transition matrix; see D. G. Schultz and J. L. Melsa: <u>State Functions and Linear Control Systems</u>, McGraw-Hill Book Company, Inc., 1967, pp. 140-142 for a discussion of this algorithm. The state transition matrix is computed in the form of matrix coefficients times the natural modes $\exp(\lambda_i t)$ so that

$$\underset{\sim}{\Phi}(t) = \underset{\sim}{F}_1 e^{\lambda_1 t} + \underset{\sim}{F}_2 e^{\lambda_2 t} + \ldots + \underset{\sim}{F}_N e^{\lambda_N t}$$

The matrix coefficients F_i are not returned in the argument list but are stored in the array CHI(I, J, K) = F_I(J, K).

   (2)   If the state transition matrix is printed by STMST by setting IKNOW = 0, then the modes associated with complex eigenvalues are written as damped sine and cosine terms. See BASMAT discussion of Sec. 2.2-3 and 2.2-4 for further details.

   (3)   The eigenvalues of the matrix A must be supplied to STMST. Complex eigenvalues must be brought into this subroutine in pair with the root with the negative imaginary part first. If CHREQ and PROOT are used to obtain the eigenvalues, they will be ordered properly. Eigenvalues must be simple.

A.20-4   *SUBPROGRAMS USED*   None.

A.20-5   *LISTING*

```
      SUBROUTINE STMST (N,A,EIGR,EIGI,IKNOW)
C  THIS SUBROUTINE DETERMINES THE STATE TRANSITION MATRIX USING
C     SYLVESTER'S EXPANSION THEOREM.
      COMMON CHI(10,10,10)
      DIMENSION A(10,10),EIGR(10),EIGI(10),SPS(10,10)
      COMPLEX CA(10,10),CA1(10,10),CA2(10,10),TCA(10,10),
     *  DENOM(10),CEIG(10)
 1000 FORMAT (1H0,5X,25HTHE ELEMENTS OF THE STATE      ,
     *      18H TRANSITION MATRIX     )
 1001 FORMAT(1H0,   5X,25HTHE MATRIX COEFFICIENT OF       ,
     *  5H EXP(,1PE13.6,7H)T*COS(,1PE13.6,2H)T/)
 1002 FORMAT (1P6E20.7)
 1003 FORMAT(1H0,   5X,25HTHE MATRIX COEFFICIENT OF       ,
     *  5H EXP(,1PE13.6,7H)T*SIN(,1PE13.6,2H)T/)
 1004 FORMAT(1H0,   5X,25HTHE MATRIX COEFFICIENT OF       ,
     *                5H EXP(,1PE13.6, 2H)T/)
```

```
1005 FORMAT (1H0,45(1H*))
     IF(IKNOW.NE.0) GO TO 800
     PRINT 1005
 800 DO 10 K=1,N
     CEIG(K)=CMPLX(EIGR(K),EIGI(K))
     DO 10 L=1,N
  10 CA(K,L) = CMPLX(A(K,L),0.0)
     I=1
     IF(IKNOW.NE.0) GO TO 700
     PRINT 1000
 700 DO 15 K=1,N
  15 DENOM(K)=CEIG(I)-CEIG(K)
     DO 500 J=1,N
     IF (J-I) 100,500,200
 100 IF (J-1) 110,110,150
 200 IF (I-1) 300,300,400
 300 IF (J-I-1) 110,110,150
 400 IF (J-I-1) 110,150,150
 110 DO 5 K=1,N
     DO 5 L=1,N
   5 CA1(K,L)=CA(K,L)
     DO 20 K=1,N
     CA1(K,K)=CA(K,K)-CEIG(J)
     DO 20 L=1,N
  20 CA1(K,L)=CA1(K,L)/DENOM(J)
     GO TO 500
 150 DO 40 K=1,N
     DO 40 L=1,N
  40 CA2(K,L)=CA(K,L)
     DO 25 K=1,N
     CA2(K,K)=CA(K,K)-CEIG(J)
     DO 25 L=1,N
  25 CA2(K,L)=CA2(K,L)/DENOM(J)
     DO 30 K=1,N
     DO 30 L=1,N
     TCA(K,L)=(0.0,0.0)
     DO 30 M=1,N
  30 TCA(K,L)=TCA(K,L) + CA1(K,M)*CA2(M,L)
     DO 35 K=1,N
     DO 35 L=1,N
  35 CA1(K,L)=TCA(K,L)
 500 CONTINUE
     IF (AIMAG(CEIG(I))) 45,50,45
  45 IM=I
     I=I+1
     IF(IKNOW.NE.0) GO TO 801
     PRINT 1001, EIGR(I),EIGI(I)
 801 DO 65 K=1,N
     DO 65 L=1,N
  65 SPS(K,L)= REAL(CA1(K,L))*2.0
     DO 66 K=1,N
     DO 66L=1,N
     CHI(IM,K,L)=SPS(K,L)
  66 CONTINUE
     IF(IKNOW.NE.0) GO TO 802
     DO 80 K=1,N
  80 PRINT 1002, (SPS(K,L),L=1,N)
     PRINT 1003, EIGR(I),EIGI(I)
 802 DO 55 K=1,N
     DO 55 L=1,N
  55 SPS(K,L)=AIMAG(CA1(K,L))*2.0
     DO 56 K=1,N
     DO 56 L=1,N
```

```
      CHI(I  ,K,L)=SPS(K,L)
 56 CONTINUE
      IF(IKNOW.NE.0) GO TO 600
      DO 85 K=1,N
 85 PRINT 1002, (SPS(K,L),L=1,N)
      GO TO 600
 50 IF(IKNOW.NE.0) GO TO 804
      PRINT 1004, EIGR(I)
804 DO 60 K=1,N
      DO 60 L=1,N
 60 SPS(K,L)=REAL (CA1(K,L))
      DO 61 K=1,N
      DO 61 L=1,N
      CHI(I,K,L)=SPS(K,L)
 61 CONTINUE
      IF(IKNOW.NE.0) GO TO 600
      DO 75 K=1,N
 75 PRINT 1002, (SPS(K,L),L=1,N)
600 IF (I.GE.N) RETURN
      I=I+1
      GO TO 700
      END
```

A.21 SUBROUTINE SUBP

The subroutine SUBP is used to form the difference of two polynomials as

$$Z(s) = X(s) - Y(s)$$

A.21-1 *CALLING SEQUENCE*

CALL SUBP (Z, IZ, X, IXA, Y, IYA)

A.21-2 *DEFINITION OF SYMBOLS*

Z = Resulting polynomial array, Z(1) is the constant term
IZ = Order of Z(s) $\leq$ 9
X = Coefficient array for X(s)
IXA = Order of X(s)
Y = Coefficient array for Y(s)
IYA = Order of Y(s)

A.21-3 *SUBPROGRAMS USED*   None.

A.21-4 *LISTING*

```
      SUBROUTINE SUBP(Z,IZ,X,IXA,Y,IYA)
C     THIS SUBROUTINE FORMS DIFFERENCE OF TWO PCLYNOMIALS
      DIMENSION X(10), Y(10), Z(10)
      IX=IXA+1
      IY=IYA+1
      ND = IX
      IF (IX -IY) 10,20,20
   10 ND = IY
   20 IF(ND) 90,90,30
   30 DO 80 I =1,ND
      IF(I-IX) 40,40,60
   40 IF(I-IY) 50,50,70
   50 Z(I) = X(I) -Y(I)
      GO TO 80
   60 Z(I) = -Y(I)
      GO TO  80
   70 Z(I) = X(I)
   80 CONTINUE
   90 IZ = ND -1
      RETURN
      END
```

A.22 SUBROUTINE TRESP

The subroutine TRESP is one of a package of four subroutines (CALCU, RUNGE, TRESP, and YDOT) used to compute the time response of a closed-loop linear system by the use of a fourth-order Runge-Kutta integration procedure. This subroutine is the master subroutine which calls the other three subprograms and also arranges the plotting of the desired variables.

A.22-1 *CALLING SEQUENCE*

Call TRESP (A, Y, B, AK, X, XMAX, H, IFREQ, N, GAIN, C)

A.22-2 *DEFINITION OF SYMBOLS*

A = Matrix array for A
Y = State vector array, $x(t)$
B = Control vector array, b
AK = Feedback coefficients, k
X = Initial time
XMAX = Final time
H = Integration step size
IFREQ = Frequency of output
N = System order, $N \leq 10$
GAIN = Controller gain K
C = Output vector array, c

A.22-3 *NOTES*

(1) This subroutine uses a fourth order Runge-Kutta algorithm (RUNGE) to integrate the linear system

$$\dot{x}(t) = A x(t) + b u(t)$$

$$u(t) = K[r(t) - k^T x(t)]$$

$$y(t) = c^T x(t)$$

from the initial time t = X to the final time t = XMAX. The integration uses a the initial condition for $x(t)$ the values of $x(t)$ passed into the subroutine as Y. The integration step size is H and the output is printed every IFREQ steps. This subroutine forms the heart of the GTRESP program; for more detail on TRESP see Sec. 2.4.

(2) This subroutine can plot the time response of the variables u(t), r(t), y(t), e(t) = r(t) - y(t) and $x_i(t)$, i = 1, 2, ... , N. The number of variables to be plotted is placed in IPLOT $\leq$ 8. Whice variables are to be plotted is indicated by placing integers in the IVAR array using the following equivalence table

1 = $x_1(t)$	8 = $x_8(t)$
2 = $x_2(t)$	9 = $x_9(t)$
3 = x_3	10 = $x_{10}(t)$
4 = $x_4(t)$	11 = e(t)
5 = $x_5(t)$	12 = u(t)
6 = $x_6(t)$	13 = y(t)
7 = $x_7(t)$	14 = r(t)

The variable IPLOT and IVAR are entered into TRESP by the use of a common statement rather than the argument list. If a plot is desired, then the number

of points to be plotted cannot exceed 101; this restriction could be removed
by redimensioning the array SKJ.

A.22-4 *SUBPROGRAMS USED*

    (1) CALCU

    (2) RUNGE

    (3) YDOT

    (4) Y8VSX

A.22-5 *LISTING*

```
      SUBROUTINE TRESP(A,Y,B,AK,X,XMAX,H,IFREQ,N,GAIN,C)
C     THIS SUBROUTINE COMPUTES AND PLOTS TIME RESPONSE
C     USING CALCU, RUNGE, YDOT AND Y8VSX
      INTEGER CHAR(15)
      COMMON IPLOT,IVAR(10)
      DIMENSION SKJ(101,9),C(10)
      DIMENSION FN(10),        Y(10),A(10,10),B(10),AK(10)
      DATA CHAR(1),CHAR(2),CHAR(3),CHAR(4),CHAR(5),
     *CHAR(6),CHAR(7),CHAR(8),CHAR(9),CHAR(10),CHAR(11),
     *CHAR(12),CHAR(13),CHAR(14),CHAR(15)/1H1,1H2,1H3,1H4,
     *1H5,1H6,1H7,1H8,1H9,1HA,1HE,1HU,1HY,1HR,1H /
   24 FORMAT (2F10.0,2I10)
   25 FORMAT (8F10.0)
   28 FORMAT (//,8X,1HT,12X,4HY(T),10X,4HU(T),4X,
     *    7(5X,1HX,I1,4H(T) ,3X))
   29 FORMAT(10(1PE14.6))
 1000 FORMAT(/,5X,33HMAXIMUM NUMBER OF POINTS EXCEEDED /)
      PRINT 28,(J,J=1,N)
      II=0
      J=0
      KOUNT=IFREQ
  300 CALL CALCU(Y,U,X,N,AK,GAIN,R)
      KOUNT = KOUNT + 1
      IF (KOUNT - IFREQ)   50, 350, 350
  350 KOUNT = 0
  450 P1=0.0
      DO 451 I=1,N
  451 P1=P1+C(I)*Y(I)
      PRINT 29,X,P1,U,(Y(M),M=1,N)
      IF (IPLOT.EQ.0) GO TO 21
      J=J+1
      IF(J.GT.101) GO TO 222
      SKJ(J,1)=X
      DO 40 I=1,IPLOT
      MM=IVAR(I)
      IF(MM.EQ.0) GO TO 40
      IF(MM.GT.10) GO TO 35
      SKJ(J,I+1)=Y(MM)
      GO TO 40
   35 KNOW=MM-10
      GO TO (36,37,38,39), KNOW
   36 SKJ(J,I+1)=R-P1
      GO TO 40
   37 SKJ(J,I+1)=U
      GO TO 40
   38 SKJ(J,I+1)=P1
      GO TO 40
   39 SKJ(J,I+1)=R
   40 CONTINUE
```

```
   21 CONTINUE
   50 CALL RUNGE (N, FN, H, X, Y, L,II)
      IF(L-1) 100, 200, 100
  200 CALL CALCU(Y,U,X,N,AK,GAIN,R)
      CALL YDOT(A,Y,FN,B,U,N)
  550 GO TO 50
  222 PRINT 1000
      GO TO 400
  100 IF(X-XMAX) 300, 300, 400
  400 IF(IPLOT.EQ.0) GO TO 403
      PRINT 600
  600 FORMAT (1H1,50X,15HSYSTEM RESPONSE//)
      PRINT 601
  601 FORMAT (48X,8HVARIABLE,8X,6HSYMBOL//)
      DO 608 I=1,IPLOT
      MM=IVAR(I)
      IF(MM.GT.10) GO TO 603
      PRINT 602,IVAR(I),CHAR(MM)
  602 FORMAT (51X,1HX,I2,13X,A1)
      GO TO 608
  603 KK=IVAR(I)-10
      GO TO(604,605,606,607), KK
  604 PRINT 610
      GO TO 608
  605 PRINT 611
      GO TO 608
  606 PRINT 612
      GO TO 608
  607 PRINT 613
  608 CONTINUE
  610 FORMAT (50X,5HERROR,12X,1HE)
  611 FORMAT (49X,7HCONTROL,11X,1HU)
  612 FORMAT (50X,6HOUTPUT,11X,1HY)
  613 FORMAT (50X,5HINPUT,12X,1HR)
      CALL Y8VSX(SKJ,J,IPLOT,10)
  403 RETURN
      END
```

A.23 SUBROUTINE YDOT

The subroutine YDOT is one of a package of four subroutines (CALCU, RUNGE, TRESP, and YDOT) used to compute the time response of a closed-loop linear system by the use of a fourth-order Runge-Kutta integration procedure. The YDOT subroutine is used to compute the derivative $\dot{x}(t) = \underline{A}x(t) + \underline{b}u(t)$.

A.23-1 *CALLING SEQUENCE*

Call YDOT(A, Y, YDOT, B, U, N)

A.23-2 *DEFINITION OF SYMBOLS*

A = Matrix array for $\underline{A}$
Y = State vector array, $\underline{x}(t)$
YDOT = Derivative vector array, $\dot{\underline{x}}(t)$
B = Control vector array, $\underline{b}$
U = Control input, u(t), from CALCU
N = System order, $N \leq 10$

A.23-3 *NOTES*

(1) Although this subroutine has been written for linear systems, it can be easily modified to handle nonlinear and time-varying systems of the form

$$\dot{\underline{x}}(t) = \underline{f}[\underline{x}(t), u(t), t]$$

by simply changing the coding for the computation of the YDOT array.

A.23-4 *SUBPROGRAMS USED* None.

A.23-5 *LISTING*

```
      SUBROUTINE YDOT(A,Y,XDOT,B,U,N)
C     THIS SUBROUTINE IS USED TO COMPUTE DERIVATIVES FOR RUNGE
      DIMENSION Y(10), A(10,10), B(10), XDOT(10)
      DO 2 I = 1, N
      XDOT(I) = 0.
      DO 1 J = 1, N
      XDOT(I) = XDOT(I) + A(I,J)*Y(J)
    1 CONTINUE
      XDOT(I) = XDOT(I) + B(I)*U
    2 CONTINUE
      RETURN
      END
```

A.24 SUBROUTINE LINEQ

The subroutine LINEQ is used to solve a set of simultaneous linear equations of the form

$$AX = Y \tag{1}$$

where A is an M x N matrix (not necessarily square) with $M, N \leq 30$, Y is a given constant M-vector and X is an N-vector of unknowns.

A.24-1 *CALLING SEQUENCE*

CALL LINEQ (A, M, N, Y, X, K, U)

A.24-2 *DEFINITION OF SYMBOLS*

A = M x N coefficient array
M = Number of rows of A
N = Number of columns of A
Y = Known constant vector
X = Particular solution to (1)
K = Nullity of A
U = Set of K basis vectors for the null space of A stored in the first K columns of U

A.24-3 *NOTES*

(1) LINEQ uses a diagonalization algorithm with row and column pivoting to solve Eq. (1). For details, see R. H. Pennington: <u>Introductory</u> <u>Computer</u> <u>Methods</u> <u>and</u> <u>Numerical</u> <u>Analysis</u>, The McMillan Company, 1965, Chapter 13.

(2) An equation of the form of (1) has a solution X if and only if

$$\text{rank } (A) = \text{rank } (A \vdots Y)$$

and the solution is unique if, in addition, A has rank equal to N. Since no restrictions are placed on A, the solution to Eq. (1) may: (a) not exist, (b) exist and be unique, and (c) exist and be nonunique. A consistency check is made on Eq. (1) during the diagonalization. If there is no solution, the message "EQUATIONS ARE INCONSISTENT" is printed and control returns to the calling program. If solution(s) exist to Eq. (1) they have the following form

$$X_t = X + Ua$$

where a is an arbitrary K-vector of constants.

A.24-4 *SUBPROGRAMS USED* None.

A.24-5 *LISTING*

```
      SUBROUTINE LINEQ(AA,NI,M,BB,X,K,U)
C     (NI*M)=DIMENSION(AA)
C     EQUATION FORM = AA*X=BB
C     K=NULLITY(AA)
C     U=SET OF BASIS VECTORS OF NULL SPACE OF AA
C     X = PARTICULAR SOLUTION
      DIMENSION AA(30,30),BB(30),A(30,31),X(30),ID(30),U(30,30)
 1000 FORMAT (27H EQUATIONS ARE INCONSISTENT)
      N=NI
      MM=M+1
```

```
      DO 200 I=1,N
      A(I,MM)=BB(I)
      DO 200 J=1,M
  200 A(I,J)=AA(I,J)
      K=1
      IF(N-M) 15,1,1
   15 IT=N+1
      N=M
      DO 16 I=IT,M
      DO 16 J=1,MM
   16 A(I,J)=0.0
    1 CONTINUE
      DO 21 I=1,M
   21 ID(I)=I
    2 CONTINUE
      KK=K+1
      IS=K
      IT=K
      B=ABS(A(K,K))
      DO 3 I=K,N
      DO 3 J=K,M
      IF(ABS(A(I,J))-B) 3,3,31
   31 IS=I
      IT=J
      B=ABS(A(I,J))
    3 CONTINUE
      IF(IS-K) 4,4,41
   41 DO 42 J=K,MM
      C=A(IS,J)
      A(IS,J)=A(K,J)
   42 A(K,J)=C
    4 CONTINUE
      IF(IT-K) 5,5,51
   51 IC=ID(K)
      ID(K)=ID(IT)
      ID(IT)=IC
      DO 52 I=1,N
      C=A(I,IT)
      A(I,IT)=A(I,K)
   52 A(I,K)=C
    5 CONTINUE
      IF(A(K,K)) 71,61,71
   61 KK=K
      K=K-1
      DO 62 J=KK,M
   62 A(J,J)=-1.0
      GO TO 6
   71 IF(K-N) 81,72,120
   72 A(N,MM)=A(N,MM)/A(N,N)
      GO TO 7
   81 DO 8 J=KK,MM
      A(K,J)=A(K,J)/A(K,K)
      DO 8 I=KK,N
      W=A(I,K)*A(K,J)
      A(I,J)=A(I,J)-W
      IF(ABS(A(I,J))-0.0001*ABS(W)) 82,8,8
   82 A(I,J)=0.0
    8 CONTINUE
      IF(K-M) 22,6,120
   22 K=KK
      GO TO 2
    6 CONTINUE
```

181

```
      DO 73 I=KK,N
      IF(A(I,MM)) 120,73,120
   73 CONTINUE
    7 CONTINUE
      K1=K-1
      DO 9 IS=1,K1
      I=K-IS
      II=I+1
      DO 9 IT=II,K
       DO 9 J=KK,MM
      A(I,J)=A(I,J)-A(I,IT)*A(IT,J)
    9 CONTINUE
      DO 10 I=1,M
      DO 10 J=1,M
      IF(ID(J)-I) 10,111,10
  111 X(I)=A(J,MM)
   11 CONTINUE
      IF(K-M) 101,10,101
  101 DO 102 IS=KK,M
  102 U(I,IS-K)=A(J,IS)
   10 CONTINUE
      K=M-K
      RETURN
  120 PRINT 1000
       RETURN
      END
```

A.25   SUBROUTINE CHREQA

The routine is used to determine the characteristic polynomial
$\det(s\underset{\sim}{I} - \underset{\sim}{A})$ for the matrix $\underset{\sim}{A}$.

   A.25-1 *CALLING SEQUENCE*

      CALL CHREQA(A, N, C)

   A.25-2 *DEFINITION OF SYMBOLS*

      A = Matrix array
      N = Dimension of $\underset{\sim}{A}$, N ≤ 10
      C = Coefficient array for the characteristic polynomial, C(1)
          is the coefficient of the constant term

   A.25-3 *NOTES*

      (1) The coefficients of the characteristic polynomial are
computed by the use of the principal-minor method; see J. L. Melsa:
"An Algorithm for the Analysis and Design of Linear State Variable
Feedback Systems," Proc. of Asilomar Conf. on Systems and Circuits,
Monterey, California, pp. 791-799, Nov. 1967.

   A.25-4 *SUBPROGRAMS USED*
      (1) DET

   A.25-5 *LISTING*

```
      SUBROUTINE CHREQA(A,N,C)
      DIMENSION J(11),C(11),B(10,10),A(10,10),D(300)
      NN=N+1
      DO 20 I=1,NN
   20 C(I)=0.0
      C(NN)=1.0
      DO 14 M=1,N
      K=0
      L=1
      J(1)=1
      GO TO 2
    1 J(L)=J(L)+1
    2 IF(L-M) 3,5,50
    3 MM=M-1
      DO 4 I=L,MM
      II=I+1
    4 J(II)=J(I)+1
    5 DO 10 I=1,M
      DO 10 KK=1,M
      NR=J(I)
      NC=J(KK)
   10 B(I,KK)=A(NR,NC)
      K=K+1
      D(K)=DET(B,M)
      DO 6 I=1,M
      L=M-I+1
      IF(J(L)-(N-M+L)) 1,6,50
    6 CONTINUE
      M1=N-M+1
      DO 14 I=1,K
   14 C(M1)=C(M1)+D(I)*(-1.0)**M
      RETURN
   50 PRINT 2000
```

183

```
2000 FORMAT (1H0,5X,15HERROR IN CHREQA)
     RETURN
     END
```

A.26   SUBROUTINE MULT

The subroutine MULT forms the product of two matrices.

A.26-1 *CALLING SEQUENCE*

    CALL MULT(A,B,C,N1,N2,N3)

A.26-2 *DEFINITION OF SYMBOLS*

    A = Matrix array for A
    B = Matrix array for B
    C = Matrix product AxB as computed by MULT
    N1 = Number of rows of A
    N2 = Number of columns of A
    N3 = Number of columns of B

A.26-3 *NOTES*

    The matrix A is stored in AD and B in BD. The matrix CD is
calculated as the product CD=ADxBD and is stored in C for return to the
calling program.

A.26-4 *SUBPROGRAMS USED*

    None

A.26-5 *LISTING*

```
      SUBROUTINE MULT(A,B,C,L,M,N)
      DIMENSION A(10,10),B(10,10),C(10,10),
     *    AD(10,10),BD(10,10),CD(10,10)
100   DO 108 J=1,M
      DO 104 I=1,L
104   AD(I,J)=A(I,J)
      DO 108 K=1,N
108   BD(J,K)=B(J,K)
      DO 112 I=1,L
      DO 112 J=1,N
      CD(I,J)=0.0
      DO 112 K=1,M
112   CD(I,J)=CD(I,J)+AD(I,K)*BD(K,J)
      DO 116 I=1,L
      DO 116 J=1,N
116   C(I,J)=CD(I,J)
      RETURN
      END
```

A.27   SUBROUTINE HERMIT

The subroutine HERMIT reduces a matrix to Hermite normal form.

A.27-1   *CALLING SEQUENCE*

    CALL HERMIT(AA,N,M,IRANK)

A.27-2   *DEFINITION OF SYMBOLS*

    AA = The matrix array whose Hermite normal form is desired
    N = Number of rows of AA
    M = Number of columns of AA
    IRANK = rank of AA

A.27-3   *NOTES*

This subroutine reduces the matrix AA to Hermite normal form by elementary row operations. The matrix AA is replaced by the Hermite normal form, and the rank of AA is returned in IRANK. The theory of Hermite normal forms is discussed in E. Nering, *Linear Algebra and Matrix Theory*, John Wiley and Sons, Inc., 1963.

A.27-4   *SUBPROGRAMS USED*

    None

A.27-5   *LISTING*

```
      SUBROUTINE HERMIT(AA,N,M,RANK)
      DIMENSION A(20,10),AA(20,10),JC(20)
      INTEGER RANK
      DO 100 I=1,N
      DO 100 J=1,M
  100 A(I,J)=AA(I,J)
      Z=1.0E-8
  110 JJC=1
      NMIN=N-1
  120 DO 210 I=1,NMIN
  130 DO 141 J=JJC,M
      JJ=J
  140 DO 141 K=I,N
      KJ=K
      IF( ABS(A(K,J)).GE.Z) GO TO 142
  141 CONTINUE
      GO TO 235
  142 JC(I)=JJ
      IF(KJ.EQ.I) GO TO 145
  143 DO 144 J1=JJ,M
      TEMP=A(I,J1)
      A(I,J1)=A(KJ,J1)
      A(KJ,J1)=TEMP
  144 CONTINUE
  145 TEMP=A(I,JJ)
      DO 148 JK=JJ,M
  148 A(I,JK)=A(I,JK)/TEMP
      IF(I.NE.1) GO TO 180
      LSTART=2
  162 DO 170 L=LSTART,N
      TEMP=A(L,JJ)
      IF( ABS(A(L,JJ)).LE.Z) GO TO 170
      DO 168 LL=JJ,M
```

```
168 A(L,LL)=A(L,LL)-TEMP*A(I,LL)
170 CONTINUE
    GO TO 200
180 IMIN=I-1
    DO 198 LI=1,IMIN
    TEMP=A(LI,JJ)
    IF( ABS(A(LI,JJ)).LE.Z)GO TO 198
    DO 196 LLI=JJ,M
196 A(LI,LLI)=A(LI,LLI)-A(I,LLI)*TEMP
198 CONTINUE
    LSTART=I+1
    GO TO 162
200 IF(JJ.EQ.M) GO TO 235
    JJC=JJ+1
210 CONTINUE
    DO 226 JK=JJ,M
    TEMP=A(N,JK)
    JJC=JK
    IF( ABS(A(N,JK)).LE.Z)GO TO 226
    DO 224 J1=JK,M
224 A(N,J1)=A(N,J1)/TEMP
    GO TO 228
226 CONTINUE
    GO TO 235
228 DO 234 LF=1,NMIN
    TEMP=A(LF,JJC)
    IF( ABS(A(LF,JJC)).LE.Z)GO TO 234
    DO 232 J1=JJC,M
232 A(LF,J1)=A(LF,J1)-TEMP*A(N,JJC)
234 CONTINUE
235 RANK=0
    DO 238 I=1,N
    SUM=0.0
    DO 237 J=1,M
    IF( ABS(A(I,J)).LE.Z)A(I,J)=0
    AA(I,J)=A(I,J)
237 SUM=SUM+ ABS(A(I,J))
    IF(SUM.GE.Z) RANK=RANK+1
238 CONTINUE
    RETURN
    END
```

B   GRAPHICAL DISPLAY PROGRAMS

B.1 INTRODUCTION

In this appendix, we will examine the graphical display subprograms which may be used to obtain graphical representations of program results on a line printer. These programs have been discussed and illustrated to some extent in preceding chapters with regard to specific programs such as GTRESP and RTLOC which make use of these plot routines. The purpose of this appendix is to give more detailed information concerning the actual use of the graphical display programs so that the reader may make additional use of them if desired. The programs presented here permit one to make X - Y plots, X versus time plot and log-log and semi-log plots.

The subprogram GRAPH is used to make log-log and semi-log plots. This subprogram, which is discussed in Sec. B.2, is used in the program FRESP to produce the Bode frequency response plots.

The programs SENSIT, FRESP and RTLOC all make use of the subprograms SPLIT for forming X - Y plots. The subprogram SPLIT is discussed in Sec. B.3.

The subprogram Y8VSX is used to plot up to eight variables versus time or can be used to form multiple X - Y cross plots if desired. The subprogram Y8VSX is discussed in Sec. B.4 and is used in the graphical time response program GTRESP.

Because all of these graphical display programs use the line printer, they possess a significant amount of quantization and should only be used to obtain approximate numerical results. In addition, it is almost impossible to write graphical display programs for general problems which give a pleasing representation upon the first trial. This situation might be compared to that of scaling an analog computer; it is only after the problem has been run a number of times that the proper scaling can be found.

These graphical display subprograms are written to produce plots on a line printer with the vertical axis across the page and the horizontal axis down the page. Hence as one usually reads a page the vertical axis is horizontal and the horizontal axis, vertical.

B.2 SUBROUTINE GRAPH

The subroutine GRAPH is used to obtain log-log and semi-log plots. The subroutine has automatic scaling so that a decade grid which is as square as possible is generated. This subprogram as currently written is set up to plot phase shift versus frequency on a semi-log plot and magnitude versus frequency on a log-log plot. Hence no scaling is done on the linear plot.

B.2-1 *CALLING SEQUENCE*

CALL GRAPH (A, B, D, ND, MMAX, MMIN)

B.2-2 *DEFINITION OF SYMBOLS*

A = Phase shift array
B = Magnitude array
D = Frequency array
ND = Number of points
MMAX = Maximum magnitude value in powers of 10, MMIN and MMAX = 0
       gives automatic scaling

B.2-3 *NOTES*   If both MMAX and MMIN are equal to zero, an automatic scaling procedure is applied to the magnitude values. The maximum and minimum values in the magnitude array are found MMAX and MMIN are set equal to the next larger and next smaller powers of ten respectively. If either MMAX or MMIN is nonzero, then these values are treated as the range of interest and all values outside this range are placed on the respective boundary.

Once the vertical grid size has been determined from the number of decades necessary to cover the magnitude variable, the horizontal, frequency grid size is selected to make the decade grid as square as possible.

B.2-4 *SUBPROGRAMS USED*

(1) MAXI

B.2-5 *LISTING*

```
      SUBROUTINE GRAPH(A,B,D,ND,MMAX,MMIN)
C     THIS IS A LOG-LOG OR SEMILOG PLOT ROUTINE
      INTEGER CHAR(10)
      DIMENSION COEFA(20),COEFB(20),NAME(5),NXB(105),DT(200),
     *NDD(200), A(ND),B(ND),D(ND),AT(200),BT(200),M(11),NX(105)
      DATA CHAR(1),CHAR(2),CHAR(3),CHAR(4),CHAR(5),CHAR(6),
     *CHAR(7),CHAR(8),CHAR(9),CHAR(10),NBK,NPD,NPP,NOO,NXX,
     *NII,NMIN,NPL/1H0,1H1,1H2,1H3,1H4,1H5,1H6,1H7,1H8,1H9,
     *1H ,1H.,1H*,1HO,1HX,1HI,1H-,1H+/
   11 FORMAT(     50X '    PHASE IN DEGREES' )
   64 FORMAT(  / 40X,26HAMPLITUDE IN POWERS OF TEN    )
  312 FORMAT(9X,I4,4(6X,I4),5X,I4,7X,I4,4(6X,I4))
  400 FORMAT ( //,5X,40HABSCISSA - RADIAN FREQ. IN POWERS OF TEN
  405 FORMAT (10X,103A1)
      PRINT 400
      NFLAG=0
      DO 12 I = 1, ND
      AT(I) = A(I)
      DT(I)=D(I)
   12 BT(I) = B(I)
    1 M(1) = -200
      DO 3 I = 2,11
    3 M(I) = M(I-1) + 40
   52 PRINT 11
   55 PRINT 312,(M(I),I = 1,11)
```

189

```
      DO 14 I = 1,ND
   14 AT(I) = PHNOM(AT(I))/4. + 50.
      IF(MMIN)69,68,69
   68 IF(MMAX)69,2,69
   69 NDEC=MMAX-MMIN
      M(1)=MMIN
      DO 690 I=1,NDEC
  690 M(I+1)=M(I)+1
      AMX=MMAX
      AMN=MMIN
      AMN1=AMN
      DO 691 I=1,ND
      IF(BT(I)-AMX) 692,692,693
  693 BT(I)=AMX
      GO TO 691
  692 IF(BT(I)-AMN) 694,691,691
  694 BT(I)=AMN
  691 CONTINUE
      GO TO 13
    2 AMX=BT(1)
      AMN = BT(1)
      CALL MAXI(BT,ND,AMX,AMN)
    9 ADIF = AMX - AMN
      IF (AMN) 65,66,66
   65 MMIN=AMN-1.
      GO TO 67
   66 MMIN=AMN
   67 AMN1=MMIN
  399 IF(AMX) 680,680,681
  680 MMAX=AMX
      GO TO 682
  681 MMAX=AMX+1.0
  682 NDEC=MMAX-MMIN
   13 DO 18 I=1,ND
   18 BT(I)=BT(I)-AMN1
  450 NLPD=100/NDEC
      NLINE=NDEC*NLPD
      NDCP1=NDEC+1
      NLP4=NLINE+4
      DO 19 I=1,NDCP1
   19 M(I)=I-1+MMIN
      AA=NLPD
      DO 21 I=1,ND
   21 B (I)=AA*BT(I)
  500 DMAX=D(1)
      DMIN=D(1)
      CALL MAXI(D,ND,DMAX,DMIN)
      IF(DMIN) 502,501,501
  501 MIND=DMIN
      GO TO 503
  502 MIND=DMIN-1.
  503 DMIN1=MIND
      IF (DMAX) 504,504,505
  504 MAXD=DMAX
      GO TO 506
  505 MAXD=DMAX+1.0
  506 DMAX1=MAXD
      NDECD=MAXD-MIND
      ADEC=NDECD
      ALPD=NLPD
      ALPD=ALPD*0.6
      LPDD=ALPD+0.5
      LTOT=NDECD*LPDD+1
```

```
      DO 508 I=1,ND
508 NDD(I)=0
      DO 520 I=1,ND
520 NDD(I)=(D(I)-DMIN1)*ALPD+1.5
      DO 408 I=1,104
408 NXB(I)=NBK
      DO 401 I=1,NDCP1
      JJ=I*NLPD-NLPD
      IF(M(I)) 402,403,403
403 NNN=NBK
      GO TO 404
402 NNN=NMIN
404 NXB(2+JJ)=NNN
      NNN=IABS(M(I)/10)
      IF(NNN) 406,406,407
406 NXB(3+JJ)=NBK
      GO TO 409
407 NXB(3+JJ)=CHAR(NNN+1)
409 NNN=IABS(M(I))-10*NNN
      NXB(4+JJ)=CHAR(NNN+1)
401 CONTINUE
140 CONTINUE
695 IF(NFLAG) 63,63,143
 63 DO 141 I=1,ND
141 BT(I)=AT(I)
      LDEC=NDEC
      LLPD=NLPD
      LLINE=NLINE
      LLP4=NLP4
      LDCP1=NDCP1
      NDEC=10
      NLPD=10
      NDCP1=11
      NLINE=100
      NLP4=104
      GO TO 200
143 PRINT 64
      PRINT 405,(NXB(J),J=2,104)
      DO 144 I=1,ND
144 BT(I)=B(I)
      NDEC=LDEC
      NLPD=LLPD
      NLINE=LLINE
      NLP4=LLP4
      NDCP1=LDCP1
200 MMM=MIND-1
      DO 201 I=1,105
201 NX(I)=NBK
      ICCT=0
      ISET=0
      LCCT=1
202 ICCT=ICCT+1
      DO 203 I=4,NLP4
203 NX(I)=NBK
      IF(ICCT.NE.LCCT)GO TO 206
      DO 204 I=4,NLP4
204 NX(I)=NMIN
      DO 205 I=4,NLP4,NLPD
205 NX(I)=NPL
      MMM=MMM+1
      ISET=100
      LCCT=LCCT+LPDD
      GO TO 208
```

191

```
206 DO 207 I=4,NLP4,NLPD
207 NX(I)=NPD
208 NX(4)=NII
    NX(NLP4)=NII
    DO 214 I=1,ND
    IF(NDD(I)-ICCT) 214,209,214
209 NRA=BT(I)
    IF(NRA) 210,211,211
210 NRA=0
211 IF(NRA-NLINE) 213,213,212
212 NRA=NLINE
213 NX(NRA+4)=NPP
214 CONTINUE
    IF(ISET) 215,217,215
215 PRINT 216,MMM,(NX(J),J=3,NLP4)
216 FORMAT(6X,I2,3X,102A1)
    ISET=0
    IF(ICCT-LTOT) 202,219,219
217 PRINT 218,(NX(J),J=3,NLP4)
218 FORMAT(11X,102A1)
    IF(ICCT-LTOT) 202,219,219
219 IF(NFLAG) 220,220,700
220 NFLAG=100
    GO TO 695
700 RETURN
    END
```

B.3 SUBROUTINE SPLIT

The subroutine SPLIT is used to obtain the cross plot of two variables on a line printer. The subprogram has automatic scaling, so that the horizontal and vertical grids are equal, producing a square grid on which true angles are shown.

B.3-1 *CALLING SEQUENCE*

CALL SPLIT (U1, VI, N)

B.3-2 *DEFINITION OF SYMBOLS*

U1 = X co-ordinate array
V1 = Y co-ordinate array
N = Number of points to be plotted, $N \leq 999$

B.3-3 *NOTES*   Each point on the plot is defined by placing its co-ordinates into the corresponding elements of the U1 and V1 arrays. It is not necessary for either of the co-ordinate arrays to be arranged in any order and it is not necessary that there be equal spacing between the co-ordinates. The subprogram SPLIT simply takes up to 999 co-ordinate pairs and plot the appropriate X - Y plot.

The subroutine carries out an automatic scaling of the plot by determining the range of the Y variable. This range is divided by ten to determine the grid spacing on both the horizontal and vertical axis. A sufficient number of grid sections is used on the horizontal axis to include the complete range of the X variable. This scaling procedure has a major advantage and disadvantage. Because the grid sections are defined identically on the horizontal and vertical axis, the resulting plot has a square grid and angles can be measured directly; this feature may be an advantage for root locus and Nyquist plots.

However if the ranges of the X and Y variables are greatly different, the plot can become either excessively spread out or condensed. For example if the range of X is 100 and that of Y is 1, then a plot 100 by 1 will be generated. This difficulty can be overcome by introducing artificial points into the co-ordinate arrays although small changes will be completely obscured. In the example above, if the Y axis range were changed to 100, then the entire plot would become a straight line. If X - Y plots must be made in which the ranges of X and Y are widely different (more than one order of magnitude), it may be best to make use of the subprogram Y8VSX although the horizontal and vertical scales will no longer be equal.

In both SPLIT and Y8VSX subprograms the automatic scaling feature may make the plot have very strange grid scale since no attempt is made to round the grid to an even number. This inconvenience can be removed, if desired, by introducing ficticious data points on a rerun of the problem.

B.3-4 *SUBPROGRAMS USED*   None.

B.3-5 *LISTING*

```
      SUBROUTINE SPLIT(U1,V1,N)
C     THIS IS AN X-Y PLOT ROUTINE
      DIMENSION LINE(101),CORD(11)
      DIMENSION X1(999),Y1(999),U1(999),V1(999)
       DATA IPLU,IZERO,IBLAN,IEYEN,MINUS,IAST/1H+,1HO,1H ,1HI,1H-
      KS=0
      DO1 K=1,N
      X1(K)=U1(K)
1     Y1(K)=V1(K)
C     CALCULATE YMAX AND YMIN
```

```
      YMAX=Y1(1)
      YMIN=Y1(1)
      DO 107 I=2,N
      IF(YMAX-Y1(I))104,105,105
104   YMAX=Y1(I)
105   IF(YMIN-Y1(I))107,107,106
106   YMIN=Y1(I)
107   CONTINUE
      IF(YMAX-YMIN)300,304,300
  304 YMAX=YMAX+0.1*Y1(I)
      YMIN=YMIN-.1*Y1(I)
  300 CONTINUE
C     ARRANGE THE DATA POINTS FOR INCREASING VALUES OF X
      NDM=N-1
      DO 200 I=1,NDM
      IA=I+1
      DO200 J=IA,N
      IF(X1(I)-X1(J))201,201,215
215   TEMP=X1(I)
      X1(I)=X1(J)
      X1(J)=TEMP
      TEMP=Y1(I)
      Y1(I)=Y1(J)
      Y1(J)=TEMP
201   CONTINUE
200   CONTINUE
      XMIN=X1(1)
      XMAX=X1(N)
C     THE NEXT CARD DETERMINES THE SPREAD OF THE X AXIS, WHEN
C     M IS 6 THEN THE GRID IS SQUARE
C
      M=6
      M1=M+1
      AM=M*10
      COR=(YMAX-YMIN)/10.
5     DO 93 JGA=1,10
      AQJ=JGA
      JGB=JGA+1
93    CORD(JGB)=AQJ*COR+YMIN
      CORD(1)=YMIN
      PRINT 6,(CORD(NR),NR=1,11)
    6 FORMAT (//10X,11(2X,1PE8.1))
      ZK=100.*YMIN/(YMIN-YMAX)
      ZL=  AM*XMIN/(YMIN-YMAX)
      LZ=ZL+1.5
      KZ=ZK+1.5
      ITAL=M1
      DO100 K=1,N
      Y1(K)=100.*Y1(K)/(YMAX-YMIN)+ZK+1.5
100   X1(K)=  AM*X1(K)/(YMAX-YMIN)+ZL+1.5
      K1=1
      KTAL=1
C     RESET
109   DO25 I=1,101
25    LINE(I) = IBLAN
C     DRAW MARGINS AND LONGITUDENAL GRID LINES
      DO 27 JV=1,101,10
27    LINE(JV) = IEYEN
  301 IF(ITAL-M1)38,37,38
37    DO 39 J=2,100
39    LINE(J) = MINUS
      DO 51 JZ=11,91,10
51    LINE(JZ) = IPLU
```

194

```
38      CONTINUE
C       DRAW ZERO REFERENCE
        IF(LZ-KTAL)164,165,164
165     DO166 I=1,101
 166    LINE(I) = IZERO
164     CONTINUE
        IF(KZ)301,301,302
 302    LINE(KZ) = IZERO
101     NX=XI(K1)
        IF(NX-KTAL)102,103,102
103     NY=YI(K1)
        LINE(NY) = IAST
        K1=K1+1
        GO TO 101
102     CONTINUE
C       GRAPH VARIABLES
        IF(ITAL-M1)32,33,32
33      AKS=KS
        ST=AKS*COR+XMIN
        KS=KS+1
        PRINT 30,ST,(LINE(K),K=1,101)
30      FORMAT(3X,1PE10.3,2X,101A1)
        ITAL=1
        GO TO 35
32      PRINT 31,(LINE(K),K=1,101)
31      FORMAT(15X,101A1)
35      ITAL=ITAL+1
        KTAL=KTAL+1
        IF(K1-N)109,109,108
108     RETURN
        END
```

B.4 SUBROUTINE Y8VSX

The subroutine Y8VSX may be used to plot up to eight variables versus time. Because the horizontal axis co-ordinates do not need to be ordered, it is also possible to use Y8VSX to produce a multiple cross plot of several Y variables versus a single X variable.

B.4-1 *CALLING SEQUENCE*

CALL Y8VSX (A, N, M, NGRID)

B.4-2 *DEFINITION OF SYMBOLS*

A = Co-ordinate array, see comments below
N = Number of horizontal values, $N \leq 101$
M = Number of Y variables, $M \leq 8$
NGRID = Number of horizontal points per grid line, see comments below.

B.4-3 *NOTES* The horizontal axis coordinates are in the first column of the A array, i.e. $A(J, 1)$. The remaining co-ordinate values for the M Y variables are in the next M columns of A, i.e. $A(J, I)$, $I = 2, 3, ..., M+1$. In addition to the variables in the argument list, the array IVAR is brought into the subroutine by means of a COMMON statement. In the first M elements of IVAR are integers (1-14) indicating which symbol should be used to identify each of the Y variables. The equivalence between the integers in IVAR and the identifying symbol is shown in Table B.4-1. This set of symbols was set up because of the use of Y8VSX in the program GTRESP. The reader may wish to define new symbols; this may be done by changing the DATA statements at the beginning of the subroutine.

Table B.4-1

IVAR	SYMBOL	IVAR	SYMBOL
1	1	8	8
2	2	9	9
3	3	10	A
4	4	11	E
5	5	12	U
6	6	13	Y
7	7	14	R

The subroutine Y8VSX operates in two different modes depending on where the variable NGRID is zero or positive. If NGRID = 0, it is not necessary that the horizontal co-ordinate be ordered. The subroutine automatically orders the array by the X co-ordinate $[A(J, 1)]$ and then determines the maximum vertical and horizontal range. Scaling is performed to generate ten horizontal and ten vertical grid lines. In general the scaling on the horizontal and vertical grid will be different. Since the X co-ordinate does not need to be ordered, one may use Y8VSX to obtain multiple X - Y plots.

If NGRID > 0, it is necessary that the X co-ordinates be ordered and equally spaced if a sensible plot is to be made. In this mode, a vertical grid is drawn and labeled every NGRID values of the X co-ordinates. This means that one X value will be on each line. If X is not equally spaced then a distorted grid will result. In this case no scaling is performed on the horizontal axis.

B.4-4 *SUBPROGRAMS USED* None.

B.4-5 *LISTING*

```
      SUBROUTINE Y8VSX(A,N,M,NGRID)
C     THIS SUBROUTINE PLOTS UP TO 8 VARIABLES VERSUS TIME
C     THIS IS BOTH A X-T AND X-Y PLOT ROUTINE
      COMMON IPLOT,IVAR(10)
      INTEGER CHAR(15)
      DIMENSION A(101,9),ABSCA(11),KAXIS(101),ORDIN(11),TEMPY(9)
      DATA CHAR(1),CHAR(2),CHAR(3),CHAR(4),CHAR(5),
     *CHAR(6),CHAR(7),CHAR(8),CHAR(9),CHAR(10),CHAR(11),
     *CHAR(12),CHAR(13),CHAR(14),CHAR(15)/1H1,1H2,1H3,1H4,
     *1H5,1H6,1H7,1H8,1H9,1HA,1HE,1HU,1HY,1HR,1H /
      DATA ISTAR,II,IPER,IDASH,IBLANK/1H*,1HI,1H.,1H-,1H /
  100 FORMAT (////,9X, 11(1PE10.2))
  101 FORMAT(1PE13.2,2X,101A1)
  102 FORMAT(15X,101A1)
      YMAX=A(1,2)
      YMIN=A(1,2)
      MP1=M+1
      DO 4 J=2,MP1
      DO 4 I=1,N
      IF(YMAX-A(I,J)) 1,2,2
    1 YMAX=A(I,J)
    2 IF(YMIN-A(I,J)) 4,4,3
    3 YMIN=A(I,J)
    4 CONTINUE
    5 YSHFT=YMIN*100.0/(YMAX-YMIN)
    6 NM1=N-1
      DO 8 I=1,NM1
      IP1=I+1
      DO 8 K=IP1,N
      IF(A(K,1)-A(I,1)) 7,8,8
    7 DO 88 J=1,MP1
      ATEMP=A(I,J)
      A(I,J)=A(K,J)
      A(K,J)=ATEMP
   88 CONTINUE
    8 CONTINUE
      XMIN=A(1,1)
      XMAX=A(N,1)
      ABSCA(1)=XMIN
      ABSCA(11)=XMAX
      ORDIN(1)=YMIN
      ORDIN(11)=YMAX
      DO 9 I=2,10
      Z=I-1
      ABSCA(I)=(XMAX-XMIN)*Z/10.0+XMIN
    9 ORDIN(I)=(YMAX-YMIN)*Z/10.0+YMIN
      PRINT 100,(ORDIN(J),J=1,11)
      STEPX=(XMAX-XMIN)/100.0
      KDELX=1
      KLINE=1
      LINE=1
      DO 26 IND=1,N
      IF(NGRID.EQ.0) GO TO 200
```

197

```
      KSTEP=LINE
      GO TO 201
 200  KSTEP=(A(IND,1)-XMIN)/STEPX+1.5
 201  DO 10 J=2,MP1
  10  TEMPY(J)=A(IND,J)*100.0/(YMAX-YMIN)-YSHFT
  11  IF(KLINE-LINE) 12,12,18
  12  DO 13 I=2,100
  13  KAXIS(I) = IDASH
      DO 14 I=1,101,10
  14  KAXIS(I) = ISTAR
      IF(KSTEP-LINE) 15,15,17
  15  DO 16 I=2,MP1
      K = TEMPY(I) + 1.5
      MM=IVAR(I-1)
  16  KAXIS(K) = CHAR(MM)
  17  PRINT 101,ABSCA(KDELX),(KAXIS(J),J=1,101)
      IF(NGRID.EQ.0) GO TO 202
      KLINE=KLINE+NGRID
      ABSCA(KDELX)=A(KLINE,1)
      GO TO 24
 202  KLINE=KLINE+10
      KDELX=KDELX+1
      GO TO 24
  18  DO 19 I=2,100
  19  KAXIS(I) = IBLANK
      DO 20 I=1,101,10
  20  KAXIS(I) = IPER
      IF(KSTEP-LINE) 21,21,23
  21  DO 22 I=2,MP1
      K = TEMPY(I) + 1.5
      MM=IVAR(I-1)
  22  KAXIS(K) = CHAR(MM)
  23  PRINT 102,(KAXIS(J),J=1,101)
  24  LINE=LINE+1
      IF (LINE-102) 25,25,27
  25  IF(KSTEP-LINE) 26,11,11
  26  CONTINUE
  27  RETURN
      END
```

NOTES

NOTES

NOTES

NOTES

NOTES

NOTES

07-041498-X